HYPERWAR

CONFLICT AND COMPETITION IN THE AI CENTURY

AMIR HUSAIN ▪ JOHN R. ALLEN
ROBERT O. WORK ▪ AUGUST COLE ▪ PAUL SCHARRE
BRUCE PORTER ▪ WENDY R. ANDERSON ▪ JIM TOWNSEND

SparkCognition Press
Austin

SparkCognition Press

SparkCognition
4030 W. Braker Ln., Ste. 500
Austin, TX 78759

First SparkCognition Press paperback edition October 2018

For information about special discounts for bulk purchases, please contact SparkCognition Special Sales at 1-844-205-7173 or info@sparkcognition.com

Manufactured in the United States of America

Library of Congress Cataloging-in-Publication Data has been applied for.

ISBN-13: 978-1-7325970-0-6

CONTENTS

COMPETITIONS

INNOVATIONS

OPERATIONS

1.

Hyperwar and Shifts in Global Power in the AI Century[1]

Hyperwar

On October 4, 1957, the Soviet Union launched the first artificial satellite, dubbed Sputnik, into Earth's orbit as the world watched in awe. This metal sphere, only the size of a beach ball, ended up spurring a technological "space race" between two competing adversaries and created a tension that has lingered for decades.

We now live in the cognitive age. An era where we will begin replicating – and exceeding – the prowess of the human mind in specific domains of expertise. While the implications of AI are very broad, as we head deeper into this new era, we will find that artificial intelligence combined with myriad exponential technologies will carry us inexorably toward a different form of warfare that will unfold at speeds we cannot fully anticipate. A form of warfare we call Hyperwar.

Since the dawn of time, the balance of power between belligerents has been dictated in great measure by the relative size of their armies. While knowledge of terrain, skill and technology have all been multipliers for smaller forces, quantity has always had a quality all its own. If one sets aside consideration of nuclear weapons, which allow small states such as Israel and North Korea to hold their opponents at bay, the outcomes of conventional conflicts are in great part determined by a country's ability to field a larger force, sustained over a longer period of time, the costs of which are enormous.

The arrival of artificial intelligence on the battlefield promises to change this. Autonomous systems are developing so rapidly, they will soon be able to perform many of the functions for which we have historically depended on soldiers, whether for intelligence analysis, decision support, or the delivery of lethal effects. In fact, if developers of these technologies are to be believed, their systems may even outperform human competition. Military power has for much of modern history been dominated by considerations of attrition warfare largely based on the capacity to inflict violence and damage from which one side or the other could not recover. Attrition warfare was hugely inefficient, and emphasized destructive outcomes. In the latter part of the 20th century the embrace of maneuver warfare carried us into an environment where success was measured in the capacity to impose one's will on the other side, not in competitive destruction. The crushing defeat of the Iraqi army (twice), where multi- and cross-domain combined arms maneuvers created the collapse of the enemy's capacity to resist, is the modern pinnacle of battlefield success. It reinforced the central tenet in war that speed of action in all matters has been, and will continue to be, decisive to military success… to victory.

As a consequence, the age-old calculation that measures a country's most basic military potential by estimating the number of able-bodied individuals capable of serving may no longer be a significant variable in determining the potency with which a country can project power.

And this potential shift promises to change much in how we view the emerging balance of power in this AI century.

In this chapter, we explore how the employment of AI and autonomy can change our view of the military balance of power in the coming decades. How does this revolution in military — and human — affairs change our understanding of our geopolitical future?

Six Key Implications

We feel the broad impact of large-scale use of autonomous military systems will not be insignificant. Specifically:

- Small, rich countries with great power ambition will significantly shape geopolitics.

- Asymmetric, insurgent use of autonomous technologies will disrupt, and could potentially destabilize, countries incapable of mounting a sophisticated, pre-emptive counter to the use of such technologies.

- Regional powers that are afforded protection by their own nuclear umbrellas will have greater freedom of action against rivals without crossing nuclear thresholds.

- Peer competitors with large, aging conventional forces will be able to amplify their conventional potential by rapidly and cheaply modernizing these assets. In many cases it will be possible to leap-frog newer technology by integrating autonomy into older platforms.

- The assurance and expectation of privacy may collapse in many parts of the world. AI applied to decision support and intelligence analysis will play a significant role in collapsing the decision-action loop. It will make ever larger data gathering efforts viable, practical, and beneficial to intelligence services. Since these activities will take place even at times when there are no formal hostilities, the question of whether there can be an expectation and assurance of privacy in the cognitive era will become paramount. And where automatically interpreted data begins to be acted upon without human intervention, entirely new questions will be raised.

- An increased level of risk tolerance to the loss of autonomous platforms and technological advances to amplify their precision should lead to lower non-combatant casualties. But this may also make conventional conflicts easier to sell in political terms.

Small, Rich... and Powerful

Recent conflicts in the Middle East have shown that countries such as the U.A.E. and Qatar, while small in absolute terms, can have an outsized effect on the region. This is partially due to their employment of technology and the considerable wealth they are able to expend on such initiatives.

The coalition opposed to Qatar, for example, accuses Qatar of sowing discord and undermining the region with its leverage of information assets such as the Al-Jazeera channel. On the other hand, the U.A.E. is seen to project its military power into distant theaters, such as Libya, through the leverage of contracted forces and a small number of more technologically advanced assets.

These examples show that when resources, technological capability and political will come together, a country with a small population can become a significant influencer of regional affairs.

In the age of Hyperwar, where the speed of conflict is enormously accelerated and with autonomous weapons making human force size less critical, this impact will be multiplied many times over.

Imagine long-range autonomous combat aircraft penetrating enemy airspace and punishing rivals. Or minimally manned helicopter carrier-sized expeditionary ships equipped with a large number of autonomous aircraft ready to project power at distant points on the globe. These barely manned fleets, networked through AI-controlled sensor platforms and autonomous decision support, would be protected by autonomous undersea assets and corvette-sized craft that fend off aerial, sub-surface, and surface threats. AI-powered, robotic "dark factories" could produce these mechanically simple systems controlled with incredibly sophisticated software.

It will be the ability of rich countries to attract key talent, acquire intellectual property developed elsewhere, and invest in capital goods to create these multi-domain autonomous force capabilities that will become a huge factor in wars of the future. Not population.

Insurgent Use of Autonomy

In the Yemen, Syrian, and Afghan conflicts, small, inexpensive commercial drones have already been used as flying IEDs (Improvised Explosive Devices), as propaganda tools, and as ISR (Intelligence, Surveillance, Reconnaissance) assets. Conventional anti-aircraft capabilities and radar infrastructure are not designed to deal with a potentially large number of small craft flying at low altitudes. The use of such systems will only continue to increase, with battlefield use creating the opportunity for improved tactics and use in a growing number of scenarios. While the drones used in Yemen and Syria were not artificially intelligent themselves or part of an artificially intelligent network, there is a strong possibility that such systems will be in the future, as AI capabilities such as image recognition and autonomous flight control continue to become common.

We see this situation heading towards the potential for a small group of individuals to literally assemble or 3D "print" robotic aviation capabilities and, eventually, perhaps a mechanized ground task force to carry out a specific mission. This phenomenon of beginning with off-the-shelf, dual-use, consumer-grade items including 3D printers, inexpensive microcontrollers, and cell phones, and ending with military-grade special operations and relatively long-range air and surface capabilities, will only be magnified as time goes on.

Numerous reports indicate that improvised weapons, including unmanned aerial vehicles, are available for sale to the highest bidder on a number of underground online weapons markets. ISIS terrorists demonstrated their ability to quickly assemble and use remote-controlled aircraft as "grenade bombers" in Iraq. Today, the capabilities of these platforms are somewhat limited. Popular Science quoted Col. Brett Sylvia of a U.S. advisory mission in Iraq as saying, "They are very short-range, targeting those front-line troops from the Iraqis."

Insurgent use of autonomy in improvised platforms will only continue to become a greater threat as the commercial technology they weaponize becomes less expensive, as they have a greater opportunity to experiment with such systems in the battlefield, and as open, online initiatives to develop autonomous flight control software become more well-known and more easily accessible to anyone with an internet connection. Law enforcement and military forces will need to counter these coming trends, and an entirely new view of defending public spaces in the domestic context will need to be considered.

Regional Powers Use Nuclear Umbrellas for Freedom of Action

The lesson North Korea learned from the fall of Saddam in Iraq, and the demise of other dictators before and after him, was that nuclear weapons are the ultimate insurance. If you have nuclear weapons, you can short circuit the grand alliances the West puts together to punish an incompliant regime. The North Koreans have, thus far, been proven right in this hypothesis. Despite testing missiles repeatedly and in threatening ways – overflying populated Japanese territory – and despite attacks against the U.S. with explosive language, no kinetic action was taken against them. The North Korean regime is also charged with incidents of terrorism and homicide carried out by North Korean operatives yet no definitive, retributive military action can be initiated against the regime. Notwithstanding President Trump's newfound interest in diplomacy with Kim Jong Un, many analysts are now of the opinion that the world will have to live with a nuclear North Korea. With ICBM capability that puts the entire United States in range, and a nuclear arsenal sufficient to ensure that Anti-Ballistic Missile (ABM) shields can no longer guarantee protection, North Korea cannot be treated lightly.

Given the situation in the Middle East, it is likely that if the Iran deal falls through, Iran would make a rapid move to acquire nuclear capability. U.S. intelligence assessments regarding how long it would take North Korea to acquire nuclear and ICBM capability turned out to be quite inaccurate; the Kim regime acquired both capabilities sooner than expected. Could this occur in the case of Iran too? And if so, what of Saudi Arabia, Turkey, and perhaps even the U.A.E.? We all remember King Abdullah's admonition of the West: Saudi Arabia will not be number two in a regional nuclear arms race behind the Iranians.

Assessing the trend in world affairs, it would be irresponsible to expect that further nuclear proliferation will not occur. It is likely, and to that extent, the North Korean scenario multiplied five-fold, in different parts of the world, is something to consider. Will all these countries use their nuclear umbrellas to increase their freedom of action? Probably.

So how can we characterize the nature and type of such action? The attack by unknown assailants against the Russian Hmeymim airbase in Syria serves as one example. Thirteen drones carrying what has been variously characterized as bombs or mortar rounds attempted to attack the airbase and the neighboring Russian naval base at Tartus. It is unlikely that these drones were being controlled by a single, AI-based swarming algorithm. They were likely controlled individually, though the attack was coordinated. Using a combination of ZSU-23 anti-aircraft defenses and electronic warfare systems, the Russians were able to defeat the attack. But what would have happened were a larger attack coordinated with more sophisticated AI-based control? Could conventional defenses deal with the threat?

Such kinetic attacks, of course, are not limited to aerial drones. They can also take place at sea, through the use of small boats carrying explosives which are potentially remote-controlled, or autonomously controlled.

Beyond these examples, the cyber domain is full of examples that suggest an ever-increasing range of possible actions available to parties in the future. For example, a CNN story published on December 13, 2017 disclosed that North Korea had been undertaking extensive actions to maliciously infiltrate thousands of computers globally to use their resources to mine cryptocurrencies such as Bitcoin. North Korean hackers were also involved in attacking Bitcoin exchanges to outright steal funds and had also developed ransomware that would seek payment from victims in order to unlock their data. The relative anonymity of cryptocurrencies, the ability to convert stolen computer resources to "money" through mining, and the lack of a central, regulatory entity monitoring transactions make cyber attacks a lucrative proposition for individuals, groups and nation-state hackers alike.

Closer to home, Newsweek revealed in early 2018 that in the preceding year, the FBI's Hostage Rescue Team was disrupted as it was in the midst of an operation. A number of small commercial drones were used to keep constant watch on the team.

These examples of low-cost cyber and kinetic attacks only scratch the surface of what may be expected in proxy wars of the future. As specific AI capabilities, such as image recognition, autonomous flight control, prediction of countermeasures, path planning, and efficient algorithms allowing the use of lower-cost systems for sensor fusion become more known and common, the asymmetric threat will grow.

But where AI poses a risk in opponents' hands, it can also provide for new types of defensive capabilities. For example, four authors of this book have written separately about how artificial intelligence-powered swarms can be used to find and attack North Korean mobile missile systems (John R. Allen and Michael E. O'Hanlon in National Interest) and defend against advanced Russian missile threats (Amir Husain and August Cole in Defense One).

AI to Revitalize Conventional Forces

For years, the Chinese PLAAF (People's Liberation Army Air Force) has been converting older fifties- and sixties-era jets such as the J-6 (Chinese version of the MiG-19) and J-7 (Chinese version of the MiG-21) into autonomous drones. It's not quite clear how the Chinese intend to employ these systems, and hypotheses range from using these as a "first wave" intended to saturate opposing air defenses, to more fanciful hypotheses involving higher degrees of autonomy that may even enable air-to-air combat. Regardless of where the capabilities of the current software stack that enables autonomy in these systems falls, what is clear is that a reimagining of otherwise obsolete equipment, married with upgrades involving sensors, autonomy, and even AI-based control algorithms, promises to open new avenues for near-peer states to rapidly field significant capability materialized via assets that had previously been written off.

Adding sensors and control to "dumb" systems isn't a new strategy for the U.S.A.F., which added "kits" such as the Boeing Joint Direct Attack Munition (JDAM) to unguided bombs, transforming them into smart, precise, GPS-guided munitions. These were used to great effect during the most recent Gulf War. Today, many manufacturers, such as Turkey's Turkish Aerospace Industries (TAI), develop such "bolt-on" kits to upgrade the thousands of unguided bombs in global armories.

Research work is demonstrating the value of machine learning in synthesizing data from multiple sensors to increase the detection accuracy and range of conventional systems such as sonar. Along the lines of the trend in the commercial space, where "software is eating the world," the weaponization of data and specialized machine learning algorithms can come together to enable quick, inexpensive upgrades that significantly enhance the capability of legacy systems. And because software travels fast – it can be modified, adapted, and developed quickly – we are entering an era where conventional systems can be upgraded quickly and modified and adapted without long engineering cycles. The speed with which a motivated adversary can imbue new capabilities and revitalize existing weapon systems will put pressure on the defense acquisition practices and policies that were largely crafted for an era that is now history. While innovation will always be valuable, the integration of existing capabilities carries with it ominous potential.

The need to modernize and change defense acquisition policy as a consequence of the AI revolution is not a major thrust of most of the mainstream defense thought pieces today, but perhaps it should be.

Is Assuring Privacy an Impossibility?

Recently, China launched a new big-data population monitoring system designed to evaluate citizens in terms of their loyalty to the state. This system of "social credits" will reportedly take into account every Chinese citizen's online activity, their shopping patterns, whether they took part in protests, their commercial and legal records, and much more. Citizens found to have a low social credit score will face automatically applied penalties. For example, they may be denied a loan application, or their freedom of movement in terms of the ability to board a domestic or international flight may be curtailed. For abysmally low scores, one wonders what the consequences might be. Perhaps incarceration?

Given the gargantuan amount of data that must be analyzed to implement such a holistic monitoring system for 1.4 billion people, it should surprise no one that much of the monitoring, collection, interpretation, and scoring will be automated and handled by AI algorithms. It should also not be particularly revelatory that for routine, high-volume actions that are likely to be taken by such a system – for example, denying a loan application above a certain amount – automation will be the only practical recourse. If it were anything else, the entire operation would simply become untenable.

In this Chinese social credit system, we have a large-scale example of AI automating the decision-action loop in a non-kinetic domain. Many countries around the world are interested in technology of this nature. Given China's willingness to export dual-use products such as UASs, it is likely that many other governments will soon obtain access to similar systems.

What becomes of privacy in this new era? What of the expectation of anonymity in certain contexts? With the backdrop of the Facebook, Cambridge Analytica, SenseTime, and other privacy debates one may find some hope in that society appears to be asking questions and pushing back against creeping surveillance. However, the realist may wonder whether most people will simply tire of this debate, becoming immune to their loss of privacy, while the de facto trend strongly leads us into an era where – at least in some parts of the world – nearly every expression and action is monitored and automatically analyzed.

While it can be argued that this is simply a natural progression in terms of authoritarian states using technology to amplify their hold on the population, beyond a certain point the difference becomes not just incremental, but capable of altering fundamental social dynamics. For example, there is a difference in an authoritarian state that can tap a few thousand telephone lines and simultaneously monitor a few thousand individuals, and one that can monitor everyone, all the time. In the former, there is space for dissent and alternate opinions to take root. There is the potential for change and a chance that some privacy and anonymity will exist. In other words, there is some hope for a dissenter. As this space shrinks, with always-on AI surveillance, will there be any ability to muster dissent? Any space for contrary opinions?

Perhaps most worryingly, will gamified systems such as a social score reprogram the citizenry en masse to become fundamentally more compliant than populations have ever been before?

Containing Human Loss, but Enabling Easier War

The fact that autonomy in weapon systems means that humans are no longer in harm's way is a good thing. But the worry surrounding this more "antiseptic" form of warfare is that war may become more likely as losses – of machines, not humans – become more "thinkable" and even acceptable. Regardless of which side of the debate one lands on, the fact remains that autonomy is coming to weapon systems. The benefits enabled by a tightening of the OODA (Observe Orient Decide Act) loop are so significant and can enable such benefits in a battle that they simply cannot be overlooked or ignored.

It is important to note that the nature of Hyperwar is not defined by autonomous weapons alone. Hyperwar as a concept and reality emerges from the speed of future war, which comes in nearly equal parts from near-to-actual real-time AI-powered intelligence analysis, and AI-powered decision support in command and control right up to and potentially including where the human figures into the decision-making. The fact that the string of detect, decide, and act – even when acting is a combination of human-controlled, or semi-autonomous systems – will be immeasurably accelerated by AI.

When one considers the changing world order – the strong drift toward multi-polarity and new power centers vehemently acting out on the world stage – together with the rapid advances in autonomous systems development, the clear implication is that certain kinds of campaigns may become "easier" for political and military leadership to justify. Will the consequences of these missions and campaigns always remain contained? Will the increased freedom of action that comes about when losses become more acceptable be abused to the point where we are likely to face greater conflict and escalatory reprisals? This remains to be seen. But it is, nonetheless, a likely outcome for which we must begin to prepare.

Conclusion

In order to plan for the world ahead, a holistic assessment that fuses consideration of technological and geopolitical change is necessary. By analyzing the rate and direction of advancement of exponential technologies such as AI, and considering this together with clear shifts developing in the world order, we can draw implications to which our planners and strategists must begin to respond. Unlike at any other time in history, the pace of change in the cognitive age in which we now live is considerably faster than what we experienced in the information age, or the industrial age before it. Relative advantage can shift quickly as "software eats the world" and value shifts to the ephemeral and easily replicable.

What is clear is that the best way to adapt to change is to stay nimble and to ramp up investments in learning, research and development. These investments can supply us with the knowledge and intellectual property resources that we can then begin to direct in rapid ways to create the 10X advantage Army Chief of Staff General Mark Milley has spoken of. Certainly, near-peer competitors, emerging regional powers, criminal groups, and even super-empowered individuals won't sit on the sidelines for too long. The capabilities and scenarios discussed here will materialize. The question is, will this be to our benefit, or to our peril?

2.

So This Is What It Feels Like To Be Offset: The U.S. Can No Longer Assume Technological and Military Superiority in the AI Era

As a champion of technological innovation, former Deputy Secretary of Defense Robert O. Work put artificial intelligence (AI) and autonomous systems at the center of the U.S. Third Offset Strategy during his tenure as the Pentagon's No. 2 official from 2014–2017. As the United States pursued next-generation systems such as AI and hypersonic missiles in recent years, so too did China and, to a lesser extent, Russia. The progress made by potential adversaries is faster than many had anticipated, leading Work to conclude that the United States risks being offset in the very technologies it expected to dominate. The following chapter is based on Work's speech in June at the Center for a New American Security 2018 annual conference in Washington.

The United States is not used to competing with strategic rivals on equal technological footing, but it is going to have to start learning or risk being left behind entirely as we enter the AI era.

Strategically, we are entering uncharted waters. This is a technological challenge we have never faced before, and turns the entire premise of the U.S. Third Offset Strategy – that game-changing advances in AI, robotics, and other technologies could confer conventional strategic advantage – on its head. For the first time, we must consider if it might be the United States that is the one being offset by countries such as China.

This is new ground for the U.S., which did not even face this kind of situation during the Cold War. While the Soviet Union surprised us with Sputnik and had very capable strategic weapons systems, their overall technological capability was far behind ours. And that was the crux of our Second Offset Strategy's focus on precision-guided conventional munitions, among other breakthroughs. We fundamentally knew U.S. and allied societies could out-innovate the Soviet Union and use those developments to change the battlefield forever. Looking at where China is today, we see something different. This is a technological competitor that's just as good as we are.

China is unequivocally working to outpace America in artificial intelligence for strategic advantage. This is simple to declare, but it has profound ramifications for the defense community because AI will be the determinative factor in how competitive the U.S. is as an economic and military power in the coming years. It cannot be stated enough. America cannot afford to lag behind.

At the highest levels, China's leaders have decided AI is the means by which they will leapfrog over the United States on the global stage. Chinese military officers in 1991 saw the strategic payoff for the U.S. military's first mover advantage in guided-munitions warfare during Operation Desert Storm and how America extended that huge advantage for 20 years. China wants to do the same thing with AI by using it to revolutionize the People's Liberation Army while simultaneously supercharging the entire economy. President Xi Jinping's plan to make his nation the world leader in AI by 2030 with a $150 billion domestic AI industry underscores the whole-of-nation commitment the PLA, the Communist Party, and industry have made, which is particularly important given the commercial origins of many defense-relevant machine learning advances in synthetic vision, data analysis, and software generation, among other innovations.

China is far from the only world power to recognize the growing importance of AI. It was Russian President Vladimir Putin who remarked in 2017 that "whoever leads in AI will rule the world." But it's increasingly clear that unlike any other potential rival for the United States, China has both the will and the means to make this goal happen.

Beijing's efforts have already paid off. China is beginning to rival, and in some cases even eclipse, the U.S. in metrics across the board, including the number of academic publications on AI, number of patents, and participation in major AI conferences. This research is being funneled directly into the Chinese government and military, who are actively working on applying AI to autonomous unmanned systems, surveillance and reconnaissance, simulations and training, intelligent decision-making, and more.

In fact, the U.S. needs to operate under the assumption that it might be falling behind.

Artificial intelligence is far from the only component of the Chinese strategy, which includes industrial and technological espionage among other measures, but it is the most impactful because of its ability to amplify the effectiveness of related systems and doctrine, such as the PLA's focus on system destruction warfare. What this means is that rather than focusing on sinking ships or shooting down airplanes, a key tenet of the Chinese strategy is to instead break apart the U.S. battle network, compromising communications programs and thereby thwarting U.S. military efforts on a massive scale. What makes this approach far more credible – and devastating – is the machine learning systems that will be the foundation of such complex and targeted cyber, electronic, and kinetic attacks as well as robotic and autonomous weapons.

And these are just the advances we know about. In offset strategies, we say you reveal for deterrence, and you conceal for warfighting effectiveness. So what else is being developed simultaneously behind the curtain? We have to be prepared to be totally surprised by China, be it with militarized AI or another warfighting capability. Consider the following worst-case scenario: What happens if the Chinese had an anti-torpedo defense that was 80 percent effective? They know that we have underwater superiority, but if they could stop 80 percent of torpedoes shot at one of their ships, how would that turn out for U.S. forces?

There is certainly historical precedence for such unexpected breakthroughs in the Asia-Pacific. Consider the long-lance torpedo, which was a Japanese weapon with extraordinarily long range in World War II. It totally surprised the U.S. Navy and created havoc for two years in the South Pacific. A development like the Japanese long-lance torpedo, whether known – such as Chinese hypersonic long-range missiles – or completely unexpected – such as an AI-enabled swarm offensive air capability – could turn the tide in battle. This is not abstract, either, given China's regional ambitions. President Xi has ordered the PLA to be ready to take on Taiwan militarily by 2020. It hardly seems wise to take such a specific directive as an idle threat. The concreteness of the date says a great deal about how close China is to technological parity, as well. Xi would not give that order if he thought the PLA technologically inferior.

While it is tempting to offer a prescribed list of action items, the truth is we already know what we need to do. We need to commit to a sense of urgency around AI investment for defense and face up to the fact that the Chinese are approaching technological parity with the U.S. in other areas too, including guided munitions and network warfare. We also need to reframe the way we talk about China's growing capability, including doing away with the term "near-peer." Imagining China as less than a peer runs the risk of greatly underestimating their own offset strategy. Even riskier, it leaves the impression that the U.S. has time that, in reality, we do not.

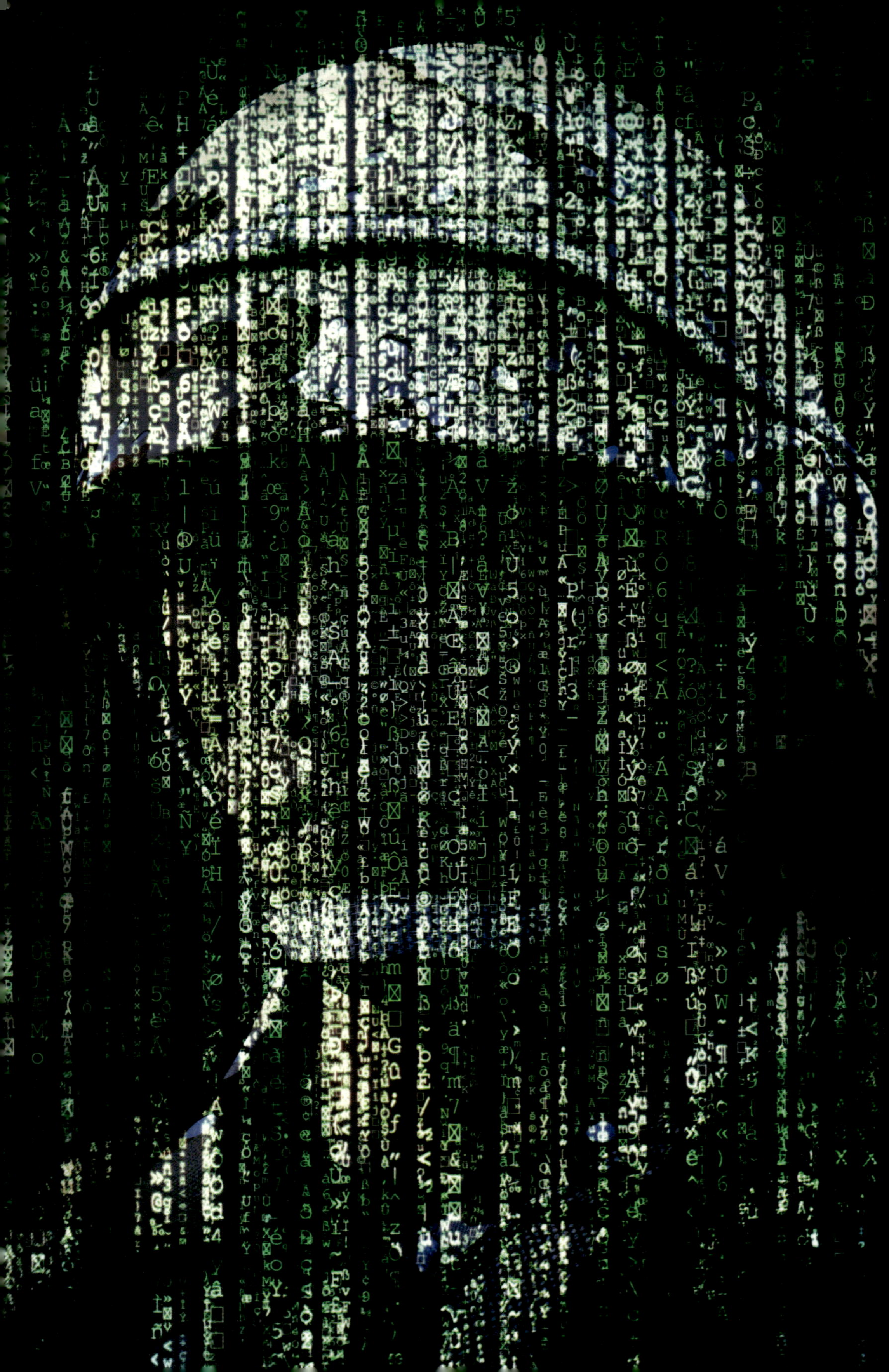

3.
The Global Race for AI Dominance

Power to Rule the World

In late August 2017, while speaking to students in Moscow, Russian President Vladimir Putin commented on the importance of artificial intelligence. He explained that "whoever becomes the leader in this sphere will become the ruler of the world." It was a direct expression of the brazen perspective to which longtime observers of the Russian President are accustomed. Was this hyperbole? Or are there valid reasons to hold the view Putin appears to hold? If so, which country has the best chance to lead the development of artificial intelligence technologies?

These are difficult and involved questions to which we will attempt an answer in this chapter. But first, what qualifies us – the authors – to take on this challenge? Our exposure to strategy, military affairs and AI is more than skin deep. John Allen commanded 150,000 troops from over 50 nations and spent 40 years living and breathing strategy, leadership, and command, while Amir Husain is a computer scientist, inventor, and the founder of one of the fastest growing AI companies in the United States. Between us, we understand weapon systems, strategies, and the dynamics of international competition, and have extensive practical experience building AI systems that utilize natural language processing, deep learning, reinforcement learning, and many other promising approaches. As a consequence, we hope our perspective will be balanced and well-rounded.

Is Putin Right?

In July 2017, we jointly published a piece in the U.S. Naval Institute's *Proceedings* explaining our view that the application of AI on the battlefield will give rise to the phenomena we call "Hyperwar," an evolution of modern warfare enabled by massively distributed autonomy and a consequent acceleration of the decision-action loop. As we've said before, this will not just be a revolution in military affairs, but in human affairs.

Artificial intelligence won't just apply to the military and the field of battle. We don't need Artificial General Intelligence – AGI, or our sci-fi vision of Commander Data from Star Trek – for this magnitude of impact. The technologies we now have and ones we are developing will suffice. Artificial Narrow Intelligence (ANI), domain specific AI capability of the type that defeated the world's leading human players of Go, is sufficient to revolutionize most areas of our society and economy.

For one, ANI systems will be used as invention machines; systems that propose new designs, ideas, and inventions in a variety of different fields, supercharging innovation and industry and leading to the creation of myriad products and capabilities that we can't imagine from our present vantage point. For another, ANI systems are already exhibiting tremendous success in automating the predictive maintenance of some of the world's most critical infrastructure across oil and gas, utilities, logistics, manufacturing, and the commercial aviation sector. Concurrently, ANI systems are also enabling autonomous control of not just cars such as Tesla, but large trucks developed by Peterbilt-Cummins and others. The degree of autonomy will continue to increase below the surface, with innovations being made by Boeing and their "autonomous submarine" Echo Voyager. This will occur on land, by military design bureaus such as Russia's Kalashnikov on the one hand, and German automaker Volkswagen on the other, as well as in the air, with systems like the Northrop Grumman X-47B and the Chinese Ehang electric passenger drone.

Eventually, we will find that every area of work and leisure – from the conversations we have with technical support agents on the phone, to the delivery of goods to our doorstep, to the diagnostic tools used by our medical professionals and the art and music we see and hear – will all leverage artificial intelligence in significant ways. AI will enable relative efficiencies and economic advantages never witnessed before.

And the important thing to remember is we don't need a sentient machine for any of this to happen. We don't need AI to do every job to automate half or more of today's national economic output.

So, Putin is right. AI is on its way to being the most potent enabler of competitive advantage.

Who Will Win?

Much like any area of scientific development, artificial intelligence requires capital and resources to develop and expand. The United States, the birthplace of artificial intelligence, has historically been both the originator of most of the important innovations in this area, and home to the largest number of top-tier research universities that have driven AI and robotics forward. Even today, few will doubt that the United States is a higher-education and research superpower. So why worry about competitors in this area? Wouldn't the U.S. be the clear leader for the foreseeable future?

Not quite.

There are many factors which make global competitors more potent than we would otherwise imagine. First, the U.S. has a very open society and research culture. Most bleeding-edge academic publications and innovations are in the public domain. The patent system is such that a filed application is published within a year of filing, and all content is available freely on the U.S. Patent Office website. U.S. universities are capturing and streaming lectures and making them available to everyone all over the world. Undoubtedly, these are excellent and commendable attributes of an open, progressive society. But they also allow competitors access to crucial ideas and research that are the products of not just years of hard work, but 200 years of societal evolution that allowed institutions like MIT, Stanford, Carnegie Mellon, and the University of Texas at Austin to exist and thrive.

Second, the U.S. may be richer overall when measured by GDP in absolute dollars (China is now richer when considering the purchasing power parity, or PPP, measure), but building things in the U.S. seems to be far more expensive than building the same thing elsewhere. There is the obvious difference in values of currency, but beyond this, there are also rules, regulations, safety concerns, and so on which can make certain kinds of development more difficult and expensive. An illustrative example is the George W. Bush-era ban on certain kinds of stem cell research. Today, China now has the world's largest sequencing capability and has conducted human gene editing via the promising CRISPR-Cas9 technique. Key, non-allied competitors to the U.S. – China and Russia – are building systems that might not be very popular in the U.S. In addition to the example of Chinese genetics research, Russian Deputy Prime Minister Dimitry Rogozin recently posted a video on Twitter showcasing an autonomous unmanned ground vehicle equipped with weapons, and a humanoid robot firing two handheld pistols.

Third, competitors seem to be in a better position to spread their technology to gain revenue, experience and influence. Soon after the 9/11 attacks, the U.S. launched the Afghan campaign. Unmanned Combat Aerial Vehicles such as the Predator and Reaper systems played a huge role in that conflict, and in the ongoing war on Al Qaeda's global network. Several U.S. allies, including Saudi Arabia, U.A.E., and Pakistan, asked to buy these same systems, but all were denied. Some sales happened, but only to a small number of allied states, most of whom have shrinking defense budgets. In contrast, Saudi Arabia is one of the biggest defense spenders in the world. As a consequence of this denial, Saudi Arabia placed the largest order on record for the Chinese CH series UAVs, complete with transfer of technology and a domestic factory to assemble these systems. Pakistan, the U.A.E. and Turkey all have non-U.S. UCAVs in their arsenal today. These UCAVs – using Chinese control software and sensors – are now being deployed in numerous conflicts, such as along the Pak-Afghan border, in Yemen, and possibly Syria. It is quite conceivable that the real-world experience in these theaters will be fed back to the manufacturer – China – in order to advance these systems. So it's not just the revenue, but also the experience… and the data, a key factor in training better algorithms. And it goes without saying that because of this policy, the U.S. has forfeit influence over employment doctrines and practices, and respective national rules of engagement.

Fourth, China has made the best possible use of the years since 9/11 when a distracted U.S. gave it all the strategic latitude required to simply focus on growth in power and capability. In the 15 years between 2001 and 2016, Chinese GDP grew more than eight-fold, from $1.3 trillion to a staggering $11.2 trillion! This money has been well used to build domestic infrastructure and fund research and education. These investments are now paying massive dividends. In 2014, China overtook the U.S. in the number of cited publications concerning the leading AI technique of "deep learning." While the number of papers alone, or citations alone, are not holistic measures, they are indicative of a broader trend. According to data gleaned from the White House's Office of Science and Technology Policy, just a year or two prior to 2014, cited Chinese papers on deep learning were less than half the number for the U.S. Beyond deep learning, China's developments in quantum communication lead the world. In June 2017, a Chinese satellite sent intertwined quantum particles from space to ground stations, over 1,200 kilometers. This form of ultra-secure, unhackable communication can provide critical advantages for diplomacy, commerce, and the military. China's investments in compute power don't lag far behind. China has a larger number of computers on the TOP500 global supercomputer ranking, and the two top global performers are both Chinese systems, TaihuLight and Tianhe 2. More secure communications, a greater amount of deep learning and AI research, and the world's most powerful computers on which to run these algorithms form a broad advantage.

Fifth, it's not just about quantity, but also the competition's quality. China has been able to attract top-end AI talent. In the past five years, Stanford's Andrew Ng was hired by Chinese firm Baidu to develop their world class AI group. MIT Technology Review recognized Baidu's Deep Speech 2 speech recognition system as one of the 10 major technology breakthroughs of 2016. The Chinese "BAT" tech trio, Baidu, Alibaba, and Tencent, are now some of the largest software companies in the world and their AI capabilities are top-class. Not only that, they are also flush with cash and able to make aggressive global acquisitions. In April 2017, Baidu announced that it was acquiring U.S.-based computer vision firm xPerception. In July 2017, they acquired an American AI chatbot company, Kitt.ai. When a state-owned Chinese investment firm, Canyon Bridge Fund, attempted to acquire the U.S.-based chip manufacturer, Lattice Semiconductor, in September, the Trump administration finally blocked this deal on national security grounds. At some stage – if not now, then in five years – as the Chinese economy continues to grow larger and at a relatively faster rate than the U.S. economy, there will come a time when such blocks and perceived "protectionist" moves will elicit an economic response. And by then, perhaps China won't need to buy U.S. companies either. The quality of their capabilities may be more broadly comparable to the U.S.

Sixth, the U.S. isn't as nimble as it used to be. The organizations we consider to be our most aggressive and nimble can't always keep up with their global competition. VentureBeat highlighted this point in an article about Andrew Ng's move to Baidu, where Ng's first move was to order Graphics Processing Unit (GPU) hardware that accelerates deep learning algorithms.

"He ordered 1,000 GPUs and got them within 24 hours," Adam Gibson, co-founder of deep-learning startup Skymind, told VentureBeat. "At Google, it would have taken him weeks or months to get that."

In our view, today the U.S. remains the global AI leader. But this lead may not last long, particularly if the fear-driven discourse in America takes hold in society and groups that seek bans on AI development extend their influence. Among others, Elon Musk has described the move to develop autonomous technologies as the equivalent of "summoning the demon." Right-wing Christian organizations, citing past work done by the authors of this chapter, have characterized AI as the "image of the beast," citing biblical verses out of context and equating AI with witchcraft. If these views, through a politicized process, result in a reduction in funding for U.S.-based AI research, or place hurdles in the way of domestic AI companies attracting capital and contracts, we will be placed at a further disadvantage. We are quite certain that China and Russia will do what they must for their national security and – just as China did with genetics and UCAVs – use any U.S.-imposed bans to do what any rational competitor would: extend their advantage and their influence.

A Cooperative Future?

In an ideal world, as humanists and scientists, we would wish for there to be a global collaboration – an international meeting of the minds – for the joint accomplishment of what could be mankind's greatest achievement: a synthetic mind. But surveying the current world scenario, analyzing the conflicts presently in process and those likely to materialize in years to come, we are not very optimistic we will see an honest, all-the-cards-on-the-table style of cooperation. In the near future, we feel governments around the world will view AI as a critical asset that enables a strategic capability that can be leveraged to create massive economic and military advantage relative to each other.

Whether we like it or not, the AI race has begun. We are in the lead today, but if present trends continue, we think the U.S. will soon be running at least second behind the Chinese. And if the Chinese engage in transfer of technology the way they have in UCAVs, we may end up possibly third.

––––––––

This chapter is based on an article that was originally published in *Foreign Policy*.

4.
On Hyperwar[1]

2 January 2018

The battle damage was devastating and constituted the leading edge of what the United States soon would discover was a widespread, strategic attack. The guided-missile destroyer had not "seen" the incoming swarm because it had not recognized that its systems were under cyber attack before things turned kinetic. The undetected cyber activity not only compromised the destroyer's sensors, but also "locked out" its defensive systems, leaving the ship almost helpless. The kinetic strikes came in waves as a complex swarm. The attack appeared to be conducted by a cloud of autonomous systems that seemed to move together with a purpose, reacting to each other and to the ship.

The speed of the attack quickly overwhelmed nearly all the ship's combat systems, and while the information technology specialists were able to release some defensive systems from the clutches of the cyber intrusion, the sailors in the combat information center (CIC) simply were unable to generate the speed to react. Decision-action times were in seconds or less. Indeed, it appeared from the now very limited situational awareness in the CIC that some of the enemy autonomous weapons were providing support to other systems to set up attacks of other systems. The entire event was over in minutes.

The captain had survived, courageously remaining on the bridge, but he was badly wounded, as were many crew members. Fires were burning out of control, and the ship was listing badly from flooding. Because of the damage, the captain was unable to communicate to the damage control assistant (DCA), who was, herself, badly wounded but valiantly seeking to control the fires and flooding. Damage control central had been hit. Evidently some of the autonomous platforms knew exactly where to strike the ship to both maximize damage and reduce the chances of survivability. With his capacity to command the ship now badly compromised and the flooding out of control, the captain did what no U.S. skipper had done for generations – he issued the order to abandon ship.

On only a few occasions has history witnessed fundamentally transformative changes in the way war is waged. The employment of cavalry, the advent of the rifled musket, and the combination of fast armor with air support and instantaneous radio communications in the execution of the Blitzkrieg strategy are a few examples. Technological developments – sometimes originating in a variety of different fields – come together to enable these seismic shifts. Another such shift is coming soon to the field of battle. Those who are not prepared for it will fare no better than the Iraqi Army did when confronted with the "Second Offset" technologies of smart, precision-guided weapons, stealth, and electronic warfare. Broad contours of how this new shift in the way war will be waged already are becoming clear. Technologies such as computer vision aided by machine learning algorithms, artificial intelligence (AI)-powered autonomous decision-making, advanced sensors, miniaturized high-powered computing capacity deployed at the "edge," high-speed networks, offensive and defensive cyber capabilities, and a host of AI-enabled techniques such as autonomous swarming and cognitive analysis of sensor data will be at the heart of this revolution. The major result of all these capabilities coming together will be an innovation warfare has never seen before: the minimization of human decision-making in the vast majority of processes traditionally required to wage war. This minimization likely will alter where the human will be located in the decision-action loop and the human's specific involvement in decision-making itself. In this coming age of Hyperwar, we will see humans providing broad, high-level inputs while machines do the planning, executing, and adapting to the reality of the mission, and take on the burden of thousands of individual decisions with no additional input.

Explaining Hyperwar

First, why refer to this AI-fueled, machine-waged conflict as Hyperwar? This is not a new term. In World War II, its use implied the global nature and many concurrent theaters of war. In today's context, however, Hyperwar may very well be applied globally, but the element of "pan-war" is not its singular defining characteristic. Instead, what makes this new form of warfare unique is the unparalleled speed enabled by automating decision-making and the concurrency of action that will become possible by leveraging artificial intelligence and machine cognition.

   In describing the wars of the future, "hyper" is used in the original Greek sense of the word – "over" or "above." This new type of combat will be beyond what has been seen before in important ways. In military terms, Hyperwar may be redefined as a type of conflict where human decision-making is almost entirely absent from the observe-orient-decide-act (OODA) loop. As a consequence, the time associated with an OODA cycle will be reduced to near-instantaneous responses. The implications of these developments are many and game-changing.

Infinite, Distributed Command & Control Capacity

Until the present time, a decision to act depended on human cognition. With autonomous decision-making, this will not be the case. While human decision-making is potent, it also has limitations in terms of speed, attention, and diligence. For example, there is a limit to how quickly humans can arrive at a decision, and there is no avoiding the "cognitive burden" of making each decision. There is a limit to how fast and how many decisions can be made before a human requires rest and replenishment to restore higher cognitive faculties.

This phenomenon has been studied in detail by psychologist Daniel Kahneman, who showed that a simple factor such as the lack of glucose could cause judges – expert decision makers – to incorrectly adjudicate appeals. Tired brains cannot carefully deliberate; instead, they revert to instinctive "fast thinking," creating the potential for error. Machines do not suffer from these limitations. And to the extent that machine intelligence is embodied as easily replicated software, often running on inexpensive hardware, it can be deployed at scales sufficient to essentially enable an infinite supply of tactical, operational, and strategic decision-making.

Concurrency of Action/Perfect Coordination

"Overpowering the enemy" is a phrase used often in the literature of war. In military terms, this refers to the concentration of force in a finite space, over a finite period of time, such that the application of this force against the opposing elements able to respond delivers a numeric or firepower advantage impossible for the opposition to counter or resist. This may not necessarily be because the attacking force is larger or more powerful than the entire defending force, only that it is more powerful when and where it matters. This is an important distinction. If a smaller force can be quickly "perfectly coordinated" and applied to a precise point where the enemy is unable to reinforce over the period of hostility, then the smaller force usually will prevail. If such action can be replicated repeatedly, then much larger opposing forces can be effectively neutralized economically and often will be dislocated psychologically.

The two key variables of concern are time and space. The time is what it takes to form and execute kinetic action, and the space is where such action is to be executed. These variables are computed as a result of significant strategic, operational, and tactical decision-making. Identifying a candidate space for the application of force is the first ingredient. When done properly, it involves computing a large set of contingencies, called branches and sequels in planning parlance, regarding the enemy's capacity to replenish, resupply, and reinforce. The tactical matters of identifying targets, maneuvering to achieve advantage or to avoid counterfire, and directing one's own fire add to this list of decisions and to the cognitive complexity. With machine-based decision-making, a large group of sensors and shooters can be coordinated instantaneously, enabling the rapid forming or massing of forces and the execution of kinetic action and subsequent dispersal.

The degree to which concurrency of action can be achieved with machine-based decision-making fuels Hyperwar and will far outpace what can be done under human control and direction.

Logistical Simplification

The old adage that "amateurs talk tactics, and professionals discuss logistics" is good guidance. Since time immemorial, waging war has required the movement of human armies that must be fed, clothed, and protected. When the level of intelligence required to fulfill a specific mission can be created in synthetic form, however, machines can become soldiers. The needs and logistical demands of robotic soldiers will not be as varied as those of a human soldier, nor will these machines be as indispensable as a human soldier. The loss of these assets no longer will trigger the expensive and dangerous standard operating procedures involving infiltration of a medical team, extraction, and transportation to a field facility.

Today's drones or unmanned combat aerial vehicles (UCAVs) mostly are remotely piloted systems that simply separate the human pilot from the craft, placing human decision-making at a distance. This is a useful configuration, but it has many downsides. First, the latencies involved mean that only certain types of missions can be fulfilled by today's drones. High-speed air-to-air combat would be difficult, for example. Second, the system remains susceptible to jamming and loss of communications. Third, the human pilot succumbs to many of the pressures and stresses of real war. This drone pilot post-traumatic stress disorder phenomenon has been well documented and sheds light on the limitations of the current model.

Truly autonomous UCAVs of a variety of types and sizes with on-board synthetic intelligence will be the foot soldiers in a future Hyperwar. Models the size of commercial quadcopters capable of weaving through forests and racing across open fields will assemble, act, and dissipate in no time. They will be armed with sophisticated sensors that feed vision and decision-making algorithms both on board, in the swarm, and when accessible, in centralized locations. In addition, they will come equipped with a variety of cyber and kinetic payloads. A large number of these systems can be coordinated by means of swarm algorithms, enabling a "collective" to ensure the fulfillment of a mission and for individual drones to support and to adapt to the loss of another.

Despite their flexibility, these systems principally will require only two resources: energy and ammunition. In the future, energy may be converted to ammunition, such as with directed-energy weapons. Still, it will be some time before the requisite miniaturization can be achieved to deliver this capability. These assets will remain "resource neutral" until they are actively being employed, reducing the overall energy required to sustain them in a theater over time. With all these changes, the logistical effort will be simplified immensely, and as a result, the "teeth to tail" ratios for autonomous forces will be higher than for any manned force.

Instant Mission Adaptations

German World War II General Erwin Rommel once said, "The best form of welfare for the troops is first-rate training." Without training, there is no chance of success, and advanced forms of military training help create specializations for roles that are essential in the conduct of war. In the face of artificially intelligent technologies and the Hyperwar they will enable, there will be two groundbreaking changes in training.

First, AI technologies such as natural language-based dialog systems that can ingest hundreds of thousands of pages of manuals, guides, studies, and more will augment human operators in non-combat situations, such as with maintenance and remediation of equipment. Eventually these capabilities will be enriched with augmented reality information-delivery technologies in combat scenarios.

Second, when employed in an entirely autonomous fashion, the tactics and strategies of an AI system – its entire set of behaviors and corpus of acquired knowledge – can be copied easily from one system to another. This is the equivalent of having the most qualified veteran instantly transfer his or her experience and expertise to troops who have never been in battle. Further, an AI system's skills and specializations can be swapped in and out immediately. The same autonomous aerial platform can be an expert "pilot" for a suppression of enemy air defenses mission and, with a quick swapping of the neural network controller, become the world's deadliest air superiority specialist. In addition, if one such "expert" AI pilot needs to be sacrificed to achieve mission objectives, so be it. Other than the hardware, nothing is lost. The "brains" of the pilot simply can be replicated on a different piece of hardware.

Training for AI-based systems can happen in the real world, or in simulators. An approach known as "reinforcement learning" has made great strides in defeating human players at traditionally unconquerable games, such as the ancient game of Go. The same technology is being employed to build better autonomous cars.

Each autonomous car does not have to go through the learning curve that every human driver must navigate. Instead, the car – or simulated car – that evolves the best performing neural network can communicate that experience and learning instantly to all other vehicles. This instant "transfer learning" will be another unparalleled reality in future Hyperwar, fueled by the employment of artificial intelligence.

28 May 2027: An Autonomous Defense Rises

The artificially intelligent cyber defense system in the guided-missile destroyer's CIC was the first to detect what appeared to be an attempt at a major cyber intrusion, perhaps an attack. The intrusion was pervasive, seeking to lock out the ship's sensors and many of its defensive systems, and seemingly concentrating on the ship's antiswarm batteries (ASBs) and supporting systems. The initial cyber attack and the successful defense occurred within microseconds. The defensive system had functioned exactly as it had been designed. As a result, the ship was able to "sense," then detect, a massive incoming complex swarm attack – the kinetic follow-up to the invisible opening strike. In fact, the system had gone further, instantly forwarding threat information to the rest of the fleet, enabling other units to prepare for an impending attack.

The captain moved quickly from the bridge into the CIC and, along with the others in the center, donned the augmented reality headgear and attendant gauntlets to assimilate and react to the totality and complexity of the battle he was about to lead. His first thought was the status of his weapons. He had only seconds as some elements of the swarm were supersonic, maybe hypersonic. Because of the elevated threat level, the captain had been given a high level of authority and autonomy to engage any potential attackers. He quickly cycled to the "weapons status" views in his headset, and all were green, being continuously fed targeting information from the ship's fire-control complex now locked onto and tracking and analyzing the incoming attacking swarm.

He had to act and shifted to the "ASB status view." With a sweep of his hand in virtual reality, he initiated the ASB.

In that instant, naval warfare changed forever. Now "cleared hot," the various components of the ASB sprinted skyward outside the skin of the ship, and the airspace was filled with several types of now-completely autonomous aerial vehicles. Some moved off at high speed on the azimuth of the incoming attack to engage the enemy swarm at long range; others dwelt in the vicinity of the ship, ready to engage as a last-ditch defense. No one on the ship, indeed no one in the U.S. Navy, had experienced the ASB going into action at full capacity. The ship shuddered as systems leapt into the air with a cacophony of noise.

Back in the CIC, the captain shifted to "target view" in his headset to see what was coming. He had been slightly skeptical this would all come together, beyond his simulator training, but now he was seeing the reality of something nearly beyond belief: completely autonomous aerial systems locked in mortal combat. Blue tracks representing ASB systems and red tracks identifying enemy threats filled the screens. Likely electronic countermeasures (ECM) decoys were highlighted in orange and automatically deprioritized by the ASB. As the battle unfolded – measured in seconds – one after another red and blue systems winked out as they crashed into each other or detonated in close proximity. That battle was moving toward his ship at a high speed. Having donned his own headset, the weapons officer quickly unleashed the full might of the various close-in weapon systems, including the autonomous systems from the ASB, which continued to engage the closing enemy swarm.

The first impact was deafening. Some elements of the enemy's swarm had detonated above the ship, taking out some of the ship's antennas. They evidently were searching for certain antennas to reduce the ship's connectivity. The second strike carried away a 20-mm. Phalanx Gatling gun, a principal means to defend the ship. The third blast struck the ship at the water line, killing and wounding a number of crew members and starting fires and flooding. While outside the ship a maelstrom was unfolding as kinetic systems autonomously coordinated fires with the near continuous launching of the ASB, inside the ship, damage-control and medical recovery measures were under way.

The captain quickly switched to "damage control view" and was able to see the AI-enabled dashboard view of the damage and the damage-control measures the ship's DCA was using to fight fires and control flooding. Because of the sophistication of the AI system, he could instantly "see" which of the ship's systems were offline, which were being rebooted to recover, and which were being instantly cross-connected to restore capacity and capability.

The AI-powered damage-control system was quickly and autonomously shifting power loads and bringing emergency systems online. Decisions for damage control were being made in seconds where before long minutes were needed.

The captain then shifted to the view he dreaded: "crew status." Because every member of the ship's company wore a "health status harness," which measured body temperature, heart rate, blood pressure, and breathing, he instantly could see the overall status of his crew and each individual sailor's status dashboard. Sobered and saddened by the number of casualties as he cycled through sub-views in this domain, he saw who had been killed and who was wounded. He knew which of his leaders were down and began to consider how he would reconstitute the chain of command.

Hours later, with his wounded cared for, the fires out, and the flooding under control, the captain reflected on the engagement. He was shaken but not frightened by the reality. The attack had come seemingly from nowhere. The cyber defense system had detected the initial cyber intrusion, and not only had it protected the ship, but it also had reasoned the attack was a precursor to something larger and alerted the CIC of what might be coming. This hypothesis had been formed, researched, and validated in less than a second. Within 10 seconds, the ship initiated general quarters on its own and the captain had donned his augmented reality ensemble. From that moment until the final fires were put out, using the automatic fire suppression system coordinated around crew status readings, the entire battle had unfolded and was over in minutes.

The autonomous nature of the ASB assets, coordinated with the CIC, and the ship's defensive systems had foiled a coordinated, complex cyber and autonomous swarm attack. The captain was struck by the realization that at nearly every point where human actions and decisions were required, they nearly risked the ship. Though he was a master of the combat systems of the U.S.S. Infinity (DDG-500), he had just experienced the near mind-numbing speeds of AI and deep learning-driven warfare. He had become the first U.S. commander to fight in the environment of Hyperwar.

Is This a Revolution in Military Affairs?

The scenarios here and the intervening discussion provide a window into only a few of the ways in which synthetic intelligence will fuel the next great shift in how warfare is conducted. The fusion of distributed machine intelligence with highly mobile platforms brings a speed and scale of concurrency never seen before. The Hyperwar these technologies will enable is a new paradigm for which we need to plan. The rise of these capabilities has sparked a revolution. But it is more than a revolution in military affairs; it is a revolution in human affairs with major implications for the security and defense arenas. Advances in AI have the capability to fundamentally change the human condition, and with it, a profoundly human undertaking, war.

Near-peer opponents already are investing heavily in these technologies and have some operational AI-powered weapon systems, such as cruise missiles. The ability for autonomous algorithms to transform moderately dangerous weapon systems into significant threats means we must watch for and guard against synthetic intelligence being added to existing arsenals.

The speed of battle at the tactical end of the warfare spectrum will accelerate enormously, collapsing the decision-action cycle to fractions of a second, giving the decisive edge to the side with the more autonomous decision-action concurrency. At the operational level, commanders will be able to "sense," "see," and engage enemy formations far more quickly by applying machine learning algorithms to the collection and analysis of huge quantities of information and directing swarms of complex, autonomous systems to simultaneously attack the enemy throughout his operational depth.

At the strategic level, the commander supported by this capacity "sees" the strategic environment through sensors operating across the entire theater. The strategic commander's capacity to ingest petabytes of information and conduct near-instantaneous analysis of information ranging from national technical means to tactical systems provides a qualitatively unsurpassed level of situational awareness and understanding heretofore unavailable to strategic commanders.

AI-powered assistive technologies – such as intelligent assistants, advanced interactive visualizations, virtual reality technologies, and real-time displays projecting rapidly updated maps – will come together to enable this situational awareness. This level of strategic understanding generates the capacity for a speed in command and control and concurrent and subsequent actions that will consistently dominate – at a time and place of our choosing – because our superior concurrency will consistently overmatch the enemy's capacity to respond.

All of this reawakens the perennial conversation about the nature and the character of war. If, indeed, we are poised at the edge of Hyperwar, we must explore the changes necessary to adapt to this new conflict environment. It will require understanding the moral dimensions of these advances, educating a new generation of leaders, and developing the AI-powered analytical systems and autonomous weapons platforms. The mental, moral, and physical challenges of Hyperwar demand analysis and a searching conversation. Our adversaries and our enemies are moving forward aggressively in this area. The United States must make the strategic investments both to be ready to wage Hyperwar and to prevent us from being surprised by it.

5.

Hyperwar: A Fictional Adversary Perspective from 2022

The following is a fictional People's Liberation Army-Navy memo from the year 2022 inspired by the opening incident depicted in the chapter "On Hyperwar." The memo is not a direct reproduction of an official PLA memo. As well, the fictitious Chinese weapons systems and military organizations depicted are meant to be illustrative of the near-future threat environment depicted in the article.

//Secret// HQ PLAN 398.09L
18 JULY 2022

FROM: RADM-C/PLAN SCITECH
TO: CNC Plan/JMC PAC/JSD-Q/TC2/TC3

REP-4/G389/001/945/OPEVAL-T

The effectiveness of the 18 JULY PLA-NAVY (PLAN) operation against DDG-69 U.S.S. MILIUS is undisputed; what follows is a necessarily rapid recounting of the incident in order to inform an urgent evaluation of the operational and technical areas where PLAN must make immediate improvements. This is imperative in order to rapidly distribute the following improvements through-out the fleet to LOON swarm units and SEA KING motherships. Our window for shoreside and underway upgrades and software revisions is shrinking, given the unexpectedly aggressive response to the incident by the United States and its allies.

Fundamentally, the decisive initial use of LOON means sur-prise is no longer on the PLAN's side; whether it was sound to em-ploy it in an initially narrow role against a single target is not my purview. But even with a limited LOON swarm and SEA KING delivery system employment, as well as combined cyber and elec-tronic attack, we have achieved a historical objective. Importantly, we have gained valuable insights heretofore unavailable in training that will yield improved operational effectiveness.

Mission Effects

A brief synopsis of the MILIUS operation that includes the ad-versary's perspective is important in order to highlight our ability to operate inside the U.S. Navy's combat systems network – and to guide our own learning. This recounting is partially informed by MILIUS CDS video, text, and audio:

0340 Local

The first LOON swarm wave (48 units) approached within strike range of the MILIUS after being delivered by a single SEA KING (S-45) submersible launch vehicle surfacing 16 kilometers ahead of MILIUS expected course. Once in range, the second and third flights were released from S-45, which then exited the area. MI-LIUS onboard defensive systems were not active because they did not detect the swarm flights due to a CDS vulnerability exploit delivered during a prior port visit in Singapore. The TAO sounded general quarters immediately after wave one struck. A fleet-wide alert message was believed to be sent from MILIUS CIC, but it was not actually transmitted.

0344 Local

The MILIUS CO reached the CIC approximately four minutes after wave one destroyed the communications and surveillance equipment, as well as the defensive systems on the ship's superstructure and deck, easing the way for waves two and three. The LOONs delivered kinetic and electronic payloads, then performed terminal flights into the MILIUS superstructure and deck.

0351 Local

LOON waves two and three (48 units each) struck approximately three minutes later, conducting further electronic attack, penetrating the superstructure, and focusing their fires on the aft VLS cells. The resulting explosions over the following four minutes compromised the hull's integrity, jammed onboard communications, and hampered damage control efforts.

0357 Local

CO MILIUS ordered the crew to abandon ship; explosions on the aft port side breached the hull in multiple locations, and MILIUS began sinking. However, due to hatch damage caused by the LOON terminal flights, the crew was unable to properly evacuate before MILIUS sank at 0359 local. No crewed PLAN surface vessels were proximate to render assistance.

Upgrade-Cycle Suggestions

The following four suggestions are based on analysis of LOON and SEA KING mission telemetry and onboard data, STARFISH coverage, and national assets:

LOON Power Management

Power management performance was not consistent throughout each wave. Battery software changes need to address peak usage during terminal-descent maneuvering, as well as on-demand energy available to power up the rocket munitions pod. Evasive maneuvers were not needed due to MILIUS CIWS being inoperable; however, if LOON units are to sortie (rearm and recharge via SEA KING), then we need to improve return-flight power. Against a CVN-class target, this is likely.

Network Resiliency

Redundant network-attack capability may be necessary if the target's network defenses detect intrusions. One option is to increase LOON flight size to create larger swarm waves. However, a better option in the MILIUS scenario would be to deploy a second SEA KING mothership with two eight-unit flights of larger SEA DRAGON aircraft. This would introduce needed ad-hoc flight heterogeneity, as well as heavier EW suppression-strike capability if CIWS or other energy counter-UAS defense systems are online. An additional SEA KING can also function as a second LOON sortie platform.

Targeting Algorithm

LOON kinetic weapons targeting performed as simulated; however, during the terminal-flight phase, LOON units predominantly struck MILIUS hatches. The LOON swarm's repeated strikes damaged hatch operation and prevented crew egress. Ultimately, it prevented MILIUS crew from abandoning ship. This was unanticipated and may have resulted from LOON machine vision interpreting MILIUS hatches as VLS targets. This is not a desired feature. Refining the machine-vision processing code for LOON flight controls should be a priority to refine terminal-flight target selection to prioritize an expanded list of ship weapons systems (VLS, main gun) to further open the targeting aperture and lessen the likelihood of confusion.

Targeted Vulnerabilities

Jamming onboard crew communications has a significant outcome on ship survivability. Hijacking MILIUS crew mobile devices and personal fitness/entertainment electronics as network nodes for onboard crew communications jamming (enabled by UNDER-TOW/PHUKET program) increased effectiveness of LOON EW and tactical cyber strikes. Extensive simulating of DDG-51 damage-control scenarios proved effective in improving LOON targeting precision, including mapping individual crew damage-control roles. Expanding shore/leave/HQ UNDERTOW crew-targeted exploits will further increase mission effectiveness. Further, UN-DERTOW exploits could be used to induce premature abandonment of ship if refinement of LOON machine-vision algorithms is not completed in time.

6.

The Intelligence Spectrum: A New Model for Evolving Human-Machine Dynamics

While definitional voids persist as to a consensus around what exactly "autonomous" means in defense policy and industry debate, we are overdue for a way to address the gap in understanding that has developed at the dawn of what can be called the "Hyperwar Era." It is time for a framework to better understand 1) the spectrum of autonomous robotic systems and 2) the dynamic relationship between human control and machine autonomy through artificial intelligence (AI). The first part is an "intelligence spectrum" model capable of encompassing advances in machine learning and AI systems based on their range of action, such as duration or scope of autonomy, and richness of perception, which includes sensors but also the ability to process data. The second part is based on the balance of human-machine interaction as well as the scope, and duration, of machine autonomy.

A New Framework

Since the MQ-1 Predator entered service more than 20 years ago, everything from its range to its datalinks have become a known commodity, even if the field of remotely operated weapons remains an emerging one. By contrast, the Russian Uran-9, a remotely-controlled robot light tank, deployed to Syria in May 2018 on its first combat assignment. It is the sort of development emblematic of the challenge faced by U.S. defense officials seeking to understand the immediate tactical ramifications as well as the strategic implications for the Pentagon's present – and future – robotic order of battle.

How then to grasp where these two military systems – one an aircraft that pioneered a new way of war, and the other a mystery whose presence portends an ominous shift in ground combat – fit in today's dynamic AI and autonomous robotics taxonomy? The civilian drone market is similarly taking off and will have as much impact on the strategic environment as purely military designs. Last year, research firm Gartner estimated the annual global drone production at 3 million units and forecast the global drone market topping $11 billion by 2020, nearly doubling from 2017.

With such growth, there is a need for a precise but scalable framework for grasping how AI and robotics will be employed not just today but into the coming decade. Today's surprise may be a robotic ground vehicle, but tomorrow's could be an autonomous network-attack algorithm capable of selecting and prosecuting its own targets that leads to a generational "before" and "after" moment. Therefore, such a framework is critical to the public- and private-sector national security and defense communities because of the worldwide commercial advances in machine learning or robotics that constantly require reassessment of current and in-development systems, as well as adversary military programs. The stakes are getting higher by the day as taxpayer money is being spent and policy is being written; there are more than 11,000 drones currently in the Defense Department inventory. Meanwhile, strategic rivals such as Russia and China pursue strategic advantages through national AI strategies.

Understanding Intelligence

The Intelligence Spectrum framework for AI and robotics begins with simple systems that have limited goals and range of action, and a narrow view of the world based on conventional sensors. It ends with broad-ranging ambitions of the sort a machine with super-intelligence might aspire to, fueled by limitless data and the ability to set its own objectives. By placing a given capability or platform on the spectrum, it is easier to see where the next-generation of technological development may be headed and how different breakthroughs, such as a new LIDAR sensor or swarm-management algorithm, alter the progression of the capability along the spectrum. Moreover, for every new weapon there is a counter-measure, and the same is true along the Intelligence Spectrum, which further reinforces the importance of understanding where an autonomous capability fits no matter how sophisticated it is.

Foundationally, simple "intelligent" systems can still be strategically important – particularly when deployed in great numbers,

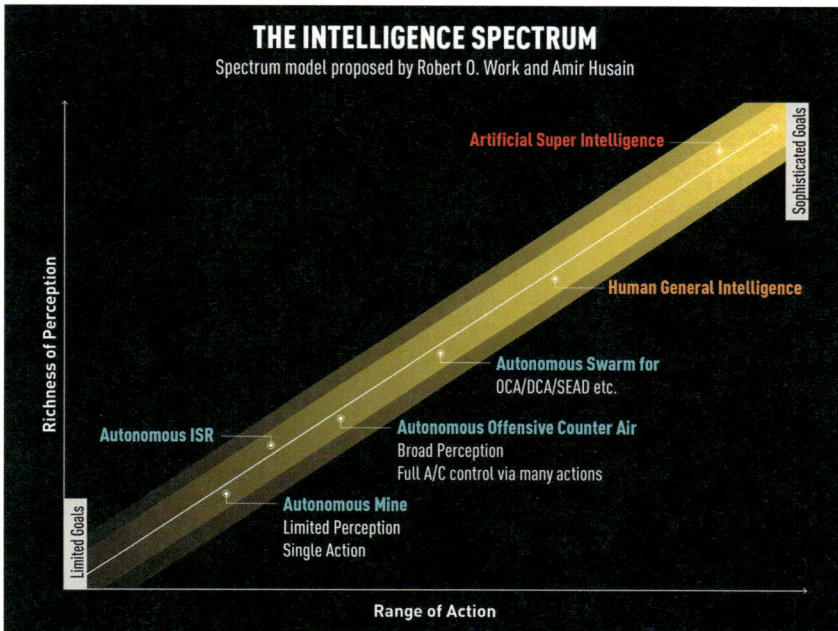

THE INTELLIGENCE SPECTRUM
Spectrum model proposed by Robert O. Work and Amir Husain

even if they do not possess swarming capabilities. An example would be an autonomous maritime mine which has a single purpose, destroying a vessel, that can deter naval freedom of movement in and along choke points like the Strait of Malacca.

The technological leap from such a mine to an undersea or airborne autonomous ISR platform is small, adding range of action defined by target sets or navigational waypoints, as well as greater sensor capabilities that enable onboard decision-making. Such a platform, like a long-duration ISR-mission glider, can still be effective at its mission even with human override or dominant control.

The relationship between effectiveness and human override becomes more elaborate, however, as range of action increases along with richness and breadth of perception. This is particularly true for missions that involve kinetic or electronic strike, such as a sortie flown by an autonomous counter-air aircraft that is targeting and releasing weapons on its own due to the machine-speed combat environment.

Further increases in perception and range of action are hallmarks of autonomous swarms used in a variety of missions, such as the aforementioned offensive counter air or suppression of enemy air defenses. This, in a sense, is a threshold expression of autonomy in which human intervention in command and control, beyond higher-order goal setting, represents an impediment to mission effectiveness.

Accordingly, from this point on, the ethical, legal, and operational dimensions become more acute for the U.S. and its allies because an even greater range of action and perception is a hallmark of human-level artificial intelligence. Beyond that capability lies artificial super intelligence, which is the supreme expression of sophisticated, and, possibly, machine-defined, goals. Though this may occur 50 or more years from today, it is an important concept to consider these other capabilities within, while also recognizing that the speed with which machine intelligence will progress is not likely to be linear, either.

How Many Humans, How Many Machines?

While Arnold Schwarzenegger may have stolen the show as the hunter-killer cyborg in 1984's "The Terminator" almost 35 years ago, the film's real star that continues to shape the AI conversation even today was the Skynet AI looming off-screen as an existential threat for humanity. In the discussion of how to understand autonomous systems, it endures as a shorthand for the extreme end of the autonomy framework: no humans, but many machines. Notably, Skynet has set higher-order goals (to exterminate humanity), it gathers resources (thinking strategically), it has unbounded autonomy (and machine armies to match), and uses kinetic weapons and other means to achieve its goals (like Schwarzenegger's time-travelling cyborg). Later films in the series even gave Skynet human form, but the unlimited and unburdened autonomy remains.

Within the autonomy framework, the duration of autonomy is a critical characteristic. It can be characterized as short, medium, long, or unlimited – as in the Skynet example. For the short- and medium-duration this could be a technical performance limitation, like fuel or battery life. But as duration increases, it is more likely to design a feature where the operational window is deliberately capped to provide a failsafe or other mission-specific characteristics. At the more extreme end of autonomy and temporal enablement this would be a system like an AI-powered cruise missile that returns to its point of departure when its "patrol" is finished. Latency, as well, represents a form of autonomy, and the ability to anchor AI-enabled sea-mines on the sea floor or bury unmanned ground vehicles until the onset of a conflict poses both a policy and operational challenge.

One of the most interesting questions today and in the near-term is how to characterize the human-machine relationship: as an expression of supervision or participation? Familiar examples would be an autonomous Black Hornet drone or autonomous tactical ISR with a kinetic payload, akin to a Switchblade drone. As the human interaction decreases, autonomy necessarily increases, which as it moves up from the tactical level to the formational level could see a transition in which a loyal wingman aircraft becomes part of an autonomous squadron with a human commander, either in the air on the ground. There is a digital manifestation of this specific paradigm of few humans and many machines, which is a highly autonomous version of the data-driven information operations performed by Cambridge Analytica in the political realm.

A threshold exists where technology bridges familiar systems like the MQ-9 Reaper to transformative autonomous ones such as

WORK-HUSAIN AUTONOMY FRAMEWORK
By Robert O. Work and Amir Husain

■ KINETIC
■ NON-KINETIC

HUMAN/MACHINE MIX	SCOPE OF AUTONOMY			
	TACTICAL/ SQUAD LEVEL	OPERATIONAL/ FORMATION LEVEL	STRATEGIC/ COMMAND LEVEL	POLITICAL
FEW HUMANS; FEW MACHINES	Autonomous tactical ISR, kinetic; example autonomous Black Hornet			
FEW HUMANS; MANY MACHINES	Loyal wingman (autonomous craft networked with human operator)	Autonomous squadron networked with human commander		Autonomous execution of influence ops (automating Cambridge Analytica type campaigns)
MANY HUMANS; FEW MACHINES	Phalanx CIWS	Autonomous network-level cyber defense		
MANY HUMANS; MANY MACHINES		Large number of autonomous embedded UGVs		
NO HUMANS; FEW MACHINES	Loitering ammunition	Autonomous ISR UAS (conventional), autonomous cyberphysical attacks	AI-powered cruise missiles	
NO HUMANS; MANY MACHINES	Kinetic hunter-killer drone swarm	Mines	Autonomous planning, targeting, and campaign execution	SKYNET: Autonomous evaluation of own interest, identification of opposition, determination of response
	SHORT TERM	MEDIUM TERM	LONG TERM	UNLIMITED
	TEMPORAL ENABLEMENT			

no humans, yet few or many machines. A current example would be loitering munition, or a hunter-killer drone swarm that uses kinetic weapons to attack targets. Moving toward greater and greater autonomy leads to autonomous attack capabilities. Moving backward in time from the point of attack, this could include autonomous planning, targeting, and campaign execution – with no human involvement. It is also possible, if not probable, that systems will have dynamic levels of autonomy that shifts according to mission objectives or strategic aims. Whether a system can shift its level of autonomy without human intervention – in a sense, a system deciding when it should give itself more operational flexibility – remains to be seen.

Conclusion

The rapid pace of change of military robotics and AI advancements is also marked by great strategic unpredictability, in part derived from surprising market-defining commercial breakthroughs that can come from literally any corner of the world. That is unlikely to change anytime soon. Therefore, we need a new framework for understanding current and emergent autonomous defense systems that can inform policy, investment, tactics, and strategy. Mapping out current systems and where they fit in offers ready insight into where capability gaps might exist or be opening up. For example, when a fully autonomous fighter might be built and deployed is not yet precisely knowable, but it is certain that it is coming. Moreover, it is also possible, and important to recognize, that such a pilotless air-superiority aircraft may not be fielded by the U.S. first. Yet it will be crucial all the same to be able to classify and understand where such a capability fits into an adversary's order of battle, as well as how it will impact U.S. and allied operations.

7.

Using AI to Enable Next-Generation Autonomy for Unmanned Aerial Systems

Abstract

This chapter covers a number of approaches that leverage artificial intelligence algorithms and techniques to aid Unmanned Combat Aerial Vehicle (UCAV) autonomy. An analysis of current approaches to autonomous control is provided followed by an exploration of how these techniques can be extended and enriched with AI techniques including artificial neural networks (ANN), ensembling and reinforcement learning (RL) to evolve control strategies for UCAVs.

Introduction

Current UAVs have limited autonomous capabilities that mainly comprise GPS waypoint following, and a few control functions such as maintenance of stability in the face of environmental factors such as wind. More recently some autonomous capabilities such as the ability for a fixed wing UCAV to land on the deck of a carrier have also been demonstrated[1]. These capabilities represent just the tip of the spear in terms of what is possible and, given both the commercial and military applications and interest, what will undoubtedly be developed in the near future. In particular, flexibility in responses that can mimic the unpredictability of human responses is one way in which autonomous systems of the future will differentiate themselves from rules-based control systems. Human-style unpredictability in action selection opens the door to finding solutions that may not have been imagined at the time the system was programmed. Additionally, this type of unpredictability in combat systems can create difficulties for adversary systems designed to act as a counter.

The capability to compute sequences of actions that do not correspond to any pre-programmed input – in other words, the ability to evolve new responses – will be another area of future differentiation. There are many other such enhancements that will be enabled via autonomous systems powered by artificial intelligence. In the following sections we will outline some of the advanced capabilities that can be engineered, and design and engineering approaches for these capabilities.

Existing Control Systems

Some degree of autonomy in flight control has existed for over 100 years, with autopilot inventor Lawrence Sperry's demonstration in 1913[5] of a control system that tied the heading and attitude indicators to a control system that hydraulically operated elevators and rudders. A fully autonomous Atlantic crossing was achieved as early as 1947 in a U.S.A.F. C-54 aircraft[6]. However, much of the early work in automating control systems were mechanical implementations of rule-based systems drawing upon cybernetics and control theory. They demonstrated that with such techniques it was possible to automate a basic mission, including takeoff and landing.

Since the 1947 demonstration, considerable effort has been invested in developing autonomous flight capabilities for commercial and military aircraft. Modern flight control or autopilot systems that govern landings are segmented in five categories from CAT-I to CAT-IIIc[11], with capabilities varying based on forward visibility and decision height. Many of these systems use rule-based, or fuzzy rule-based control, incorporating sensor-fusion techniques such as Kalman filters[7]. They are capable of following a planned route and adjusting for environmental factors such as cross-winds, turbulence and so on.

The increased popularity of commercial drones, and the heightened utilization of military drone aircraft has, in parallel, created a new class of autonomous capabilities from open source initiatives such as the ArduPilot[8] flight control software for low-cost drones to higher levels of autonomy in military drones. Software such as the ArduPilot, for example, uses a combination of GPS positioning, additional sensors to gauge velocity and position, combined with basic flight control rules to autonomously navigate to a sequence of waypoints. Many of these map input-based waypoint following capabilities are also implemented in military surveillance and combat drones.

Another area of control innovation comes from swarm theory and related control algorithms. At the simplest level, these algorithms seek inspiration from the behavior of biological systems such as ant colonies or flocks of birds. They are collaboration algorithms that enable each individual system in the swarm to compute its future actions based on its own measurements, but also those of its neighbors. While basic swarm algorithms[10] are effective in providing coverage over an area, and automatically repositioning all nodes when one is lost to maintain coverage, they do not provide much guidance on how to divide mission responsibilities and burdens, and to effectively delegate them to individual nodes. The concept of a "swarm" as found in biology will have to evolve into something entirely different – perhaps somewhat similar to a pack hunt, but even that analogy would only be marginal – in order for it to be an effective and useful system, particularly in a military context.

Some of the reasons why we propose this conclusion regarding the inadequacy of existing swarm algorithms is that most biologically inspired algorithms, such as Particle Swarm Optimization (PSO)[12] or Artificial Bee Colony Algorithm (ABC)[13], are search or optimization techniques that do not account for the role of an individual particle (or node) in the swarm. For example, PSO proposes the same meta-heuristic for computing positional updates for all points and does not incorporate a differential update mechanism based on the role of a particle. In a subsequent publication, we intend to propose a "Pack Hunt Optimization" (PHO) algorithm that we believe addresses the shortcomings of the existing swarm algorithms we have cited, and holds relevance to UCAV control applications. The state of current control systems can be summed up as follows:

- Effective at basic navigation and path following
- Many existing techniques to fuse sensor data for accurate position identification
- Able to automatically take off and land if runways are properly instrumented
- Actions beyond flight control (such as weapons engagement) are presently manual
- Missions are pre-defined
- Swarm algorithms can provide additional value for relative positioning of multiple assets and distributed sensing

Advanced Autonomous Capabilities

The purpose of this section is to outline a few areas of potential advancement that can be expected of autonomous systems of the future. This list is neither exhaustive nor complete with regards to the author's current conception of all such advanced capabilities. It is a subset of possible functions that is listed to illuminate the broad contours of what is possible in terms of applications of artificial intelligence to UCAV autonomy. Some features include:

Knowledge & Assessment Updates

- Identification of potential threats outside pre-programmed mission briefs
- Autonomous exploration and assessment of identified targets that autonomous control deems to be high priority
- Enhancement and update to intelligence supplied as part of the mission brief and plan, based on actual observation

Autonomous Navigation and Swarm Coordination

- Ability to adjust to environmental conditions that cause system or any linked swarm systems to deviate from mission plan expectations
- Ability to adjust to loss of a swarm asset, not just in terms of repositioning, but including potential re-tasking (i.e. assumption of a new role on the part of an individual asset)

Autonomous Evasion

- Automated update to mission plan based on sensor detection of probable manned aerial intercept
- Automated update to mission plan based on detection of unexpected sensor presence
- Autonomous evasion in the event of a RWR (Radar Warning Receiver) activation or MAW (Missile Approach Warning) system activation

Autonomous Targeting

- Autonomous addition to target lists based on computer vision or alternate sensor-based identification of threats to mission (including surface to air threats)
- Autonomous addition to target lists in the event that primary targets have already been neutralized
- Autonomous deletion of a target from target lists in the event it has been already neutralized, is found to violate a "hard" policy constraint, or is low priority and its neutralization harms the overall achievement or success of the mission

The Need for a New Approach

In the preceding sections we explored the current state of autonomous systems and the rules-based approach that is often employed to develop these systems. Further, we also considered a number of advanced capabilities that would be desirable in future autonomous control systems. A fundamental challenge in developing these future capabilities is that the range of scenarios an autonomous system would have to contend with in order to effectively execute the required maneuvers are enormous. Tackling such a large range of possibilities with a rules-based system will be impractical not only because of the combinatorial explosion of possibilities that would require individual rules, but also because human designers of such a system may simply not be able to conceive every imaginable scenario the autonomous system could find itself in.

Another challenge is that rules-based systems are hard coded to measure certain criteria, or sensor values, and then act based on this pre-specified criteria. This hard coding means that each rule is tied to a specific set of sensors. If additional sensors are added

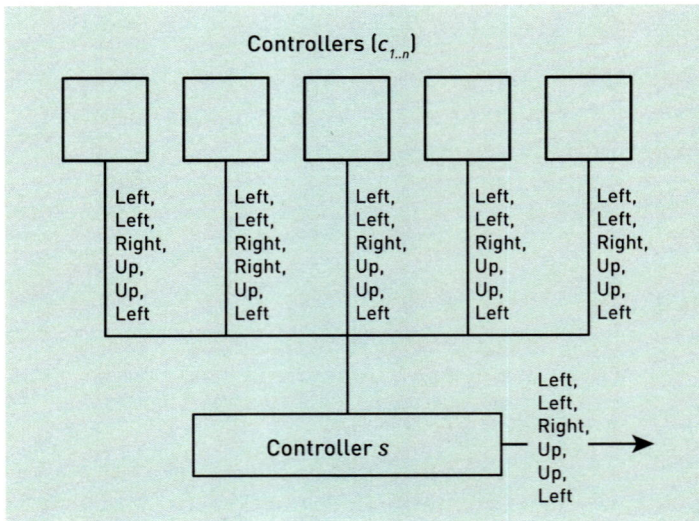

Controllers $(c_{1..n})$

Left, Left, Right, Up, Up, Left

Left, Left, Right, Right, Up, Left

Left, Left, Right, Up, Up, Left

Left, Left, Right, Up, Up, Left

Left, Left, Right, Up, Up, Left

Controller s

Left, Left, Right, Up, Up, Left

to a system, or existing sensors are upgraded, a large number of rules would have to be re-written, creating an obvious cost and effort burden.

What we have described above is far from an exhaustive list of limitations in current autonomous systems, but we believe they are sufficient to motivate the need for a new architecture for autonomy. A future system that moves beyond rules-based systems, incorporates learning capabilities so that actions can be learned rather than hard coded, and can adapt to new information from new or better sensors will represent a substantial advance. In the sections that follow, we define the contours of just such a system.

An Architecture for Advanced Autonomy

The fundamental architecture we propose in this chapter is based on multiple independent control systems connected to an action optimizer neural network. Each of the multiple independent control systems can be neural networks or non-ANN rule-based control systems that output a suggested vector of actions or control activations. The action optimizer ANN gates and weighs the inputs supplied by each independent control system.

Let c_k be an independent control system, and s be an action optimizer neural network to which $c_{1..n}$ control networks are connected. Additionally, let set E contain a collection $e_{1..m}$ of environmental inputs that are supplied to s. Then, we denote the specific configuration of all environmental inputs at time t by E_t and the output of s under these environmental inputs and based on the inputs of all independent control networks, as follows:

$$s\ (E^t,\ C^t) = A^t$$

The goal of our system is to optimize the selection of action sequences $A^{t..t+k}$ such that this sequence maximizes the performance of the system being controlled. It is important to understand what we mean by "performance" here. We define performance as a variable that is the output of a utility function U such that this output is high when the weighted achievement of all mission parameters is large, and low when the weighted achievement of mission parameters is small. In other words, we are attempting to locally maximize U:

$$\frac{dU}{dx} = 0$$

and:

$$\frac{d^2U}{dx^2} < 0$$

The question obviously arises, how do we build the function s? Conventionally, control functions have been built in various ways, for example as fuzzy rule-based systems[3]. However, we propose to implement the control function s as an artificial neural network (ANN). As the application at hand will benefit from some knowledge of past actions, we specifically propose to implement the network as a recurrent neural network (RNN).

Evolving Mission-Specific Controllers

The actual training and evolution of the RNN represented by s is not the subject of this chapter and will be documented in a subsequent publication. In summary, this can be done in a manner that combines real world and simulator environments. However, in a more detailed future exploration we intend to cover questions such as whether individual control networks, $c_{1..n}$, can be trained independently and how a training set that reflects the wide range of scenarios the UCAV might experience would be compiled. For the purpose of the present discussion, our basic approach is to use reinforcement learning (RL) techniques[4] to train the RNN in a simulated environment until a basic level of competence has been achieved, and to then allow the evolved network to control a real craft. Collected data from the actual flight is reconciled with the simulated environment and the process is repeated until an acceptable level of capability is demonstrated by s. This reconciliation would benefit from applications of transfer learning[14].

One of the benefits of this approach is that the simulated environment can introduce environmental constraints that s must respond to appropriately. For example, these can be navigation constraints such as avoiding certain pre-identified objects on a map. Work has already been done to use search algorithms such as $A*$ to find viable paths around objects to be avoided[2] and this type of constraint can be implemented by one of the independent control networks (c_k, as presented in the previous section). Other examples of existing work that could be leveraged in the form of an independent control network include collaborative mapping algorithms for multiple autonomous vehicles[15]. Of course, other constraints and optimizations would be represented by other ensembled control networks, forcing s to weight them and choose from them carefully, in a way that maximizes U.

Thus, the controller can be evolved to optimize operation in different types of environments, and under different constraints. It may then become possible to simply "upload" the optimal controller for a particular environment, or a particular mission type, into the same craft and achieve mission-specific optimal performance.

Semantic Interpretation of Sensor Data

Sensor data in autonomous systems does not have to remain limited to environmental measurements or flight sensor readings. It can include a variety of image feeds from forward, rear or down-facing cameras. Additionally, radar data and forward looking infrared (FLIR) sensor data is also a possibility. In order to utilize all this diverse data to make decisions and even deviate in small but important ways from the original mission plans, all of this data has to be interpreted and semantically modeled. In other words, its meaning and relevance to the mission and its role in governing future action has to be established.

For the purpose of understanding how such data can be interpreted and what its impact on decisions can be, we classify sensors and data sources into the following categories:

ACTIONS	HEALTH	PERF.	NAV.	ENVIR. MAPPING
Terminate Mission	X	X		X
Update Mission Achievement				X
Add New Target				X
De-prioritize Target	X			X
Change Course	X		X	X
Add Obstacle (Constrain Path)				X
Engage Weapon System				X
Evasive Maneuvers				X
Engage Countermeasures				X

Internal Sensors

- System health (e.g., engine vibration, various temperatures and internal system pressure)
- System performance (e.g., velocity, stress)

External Sensors

- Navigational aides (e.g. level, wind speed, inertial navigation gyroscopic sensors)
- Environmental mapping (e.g., camera, radar, LIDAR, FLIR, RWR, MAWS)

In the following example table, we show the types of impact that information received from these sensors can potentially have on mission plans and vehicle navigation.

In order to support the types of advanced autonomy outlined in this chapter, many of the actions highlighted in the table will likely need to be combined based on sensor input to form a chain of actions that update the internal state and maps used by the autonomous asset. Sensor data may be an input required by any controller c_k or by the controller s. Thus, a sensor bus connects all sensors to all controllers.

For many sensor types, instead of the sensor providing a raw output, we transform the output to reflect semantic constructs. For example, instead of a raw radar signal input, we may transform

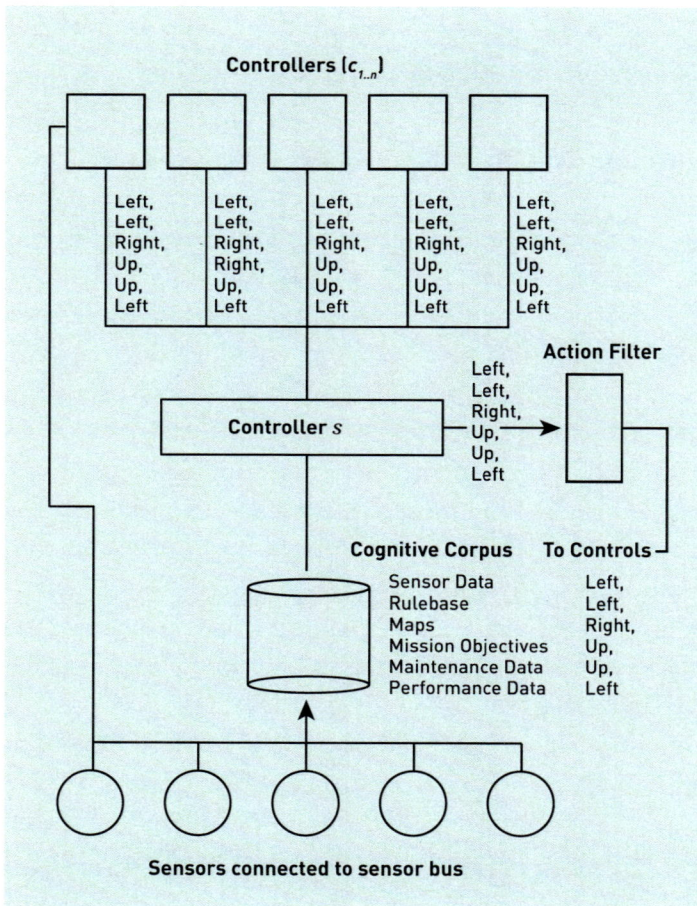

the signal into a data structure that reflects the position, speed, heading, type and classification of each detected object. This transformation of raw sensor data into semantic outputs that use a common data representation for each class of sensor enables re-placeability of underlying components so that the same controllers can work effectively even when sensors are replaced or upgraded.

The semantic output of individual sensor systems can be used by controllers, and is also stored in a cognitive corpus, which is a database that can store mission information, current status, maps, objectives, past performance data, and not-to-violate parameters for actions that are used to gate the final output of the controller s.

Knowledge Representation for Advanced Autonomy

As the more complete diagram of the proposed autonomy ar-chitecture illustrates, the controller s receives input from a set of controllers $c_{1..n}$ and is also connected to the sensor bus and the cognitive corpus. A state map stored in the cognitive corpus re-flects the full environmental picture available to the autonomous asset. For example, it includes an estimate of the asset's own posi-tion, the positions of allied assets, the positions of enemy assets, marked mission targets, paths indicating preferred trajectories at the time of mission planning, territory and locations over which to avoid flight, and other pertinent data that can assist with route planning, objective fulfillment, and obstacle avoidance.

This state map forms another important input to the controller s as it chooses the most optimal sequence of actions. The image below shows a visual representation of what the state map might track. Here, it shows the location of multiple allied assets, for ex-ample systems that might be part of a swarm with the UCAV that is maintaining this map. There is also a hostile entity identified with additional information regarding its speed and heading. Lo-cations on the ground indicate sites to be avoided. Sensor infor-mation carried in set (or vector) E results in updates to the state of each object in this map. Note that the state map is maintained

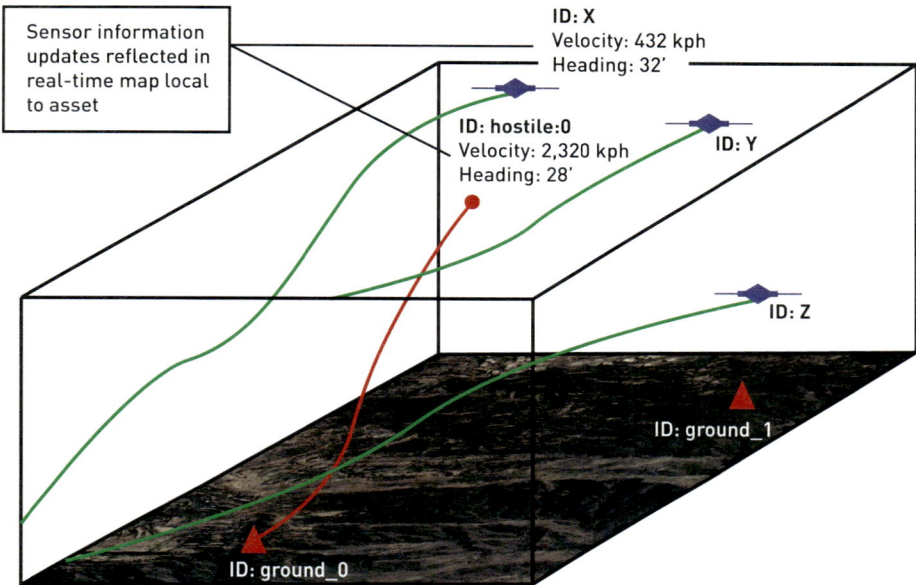

Sensor information updates reflected in real-time map local to asset

ID: X
Velocity: 432 kph
Heading: 32'

ID: hostile:0
Velocity: 2,320 kph
Heading: 28'

ID: Y

ID: Z

ID: ground_1

ID: ground_0

by each autonomous asset and, while the underlying information used to update it may be shared with or received from other systems, each autonomous asset acts based on its own internal representation, or copy, of the state map.

While the details of an implementation are beyond the scope of this chapter, we propose that the information exchange between autonomous systems occur using a blockchain protocol[9]. Benefits of this approach include the fact that in the event communication is interrupted and updates are missed, information can be reconstructed with guarantees regarding accuracy and order. Further, the use of a blockchain store ensures that a single or few malicious participants cannot impact the veracity of the information contained therein. While the figure below shows a graphic representation of the map, it is possible to represent such a map as a vector or matrix. By doing so, it can readily be supplied to the controllers as an input.

Conclusion

Sophisticated autonomy requires control over a wider range of action than rule-based systems can support. The subtle changes in flight patterns, identification of new threats, self-directed changes in mission profile, and target selection all require autonomous assets to go beyond pre-ordained instructions. Machine learning and AI techniques offer a viable way for autonomous systems to learn and evolve behaviors that go beyond their programming. Semantic information passing from sensors, via a sensor bus, to a collection of decision-making controllers provides for plug and play replacements of individual controllers. An artificial neural network such as an RNN can ensemble and combine inputs from multiple controllers to create a single, coherent control signal. In taking this approach, while some of the individual controllers may be rules-based, the RNN really evolves into the autonomous intelligence that can consider a variety of concerns and factors via control system inputs, and decide on the optimal action. We propose delinking control networks from the ensembler RNN so that individual control RNNs may be evolved and trained to execute differing mission profiles optimally, and these "personalities" may be easily uploaded into the autonomous asset with no hardware changes necessary. One of the challenges in taking this advanced approach may be the inability to guarantee what exactly a learning, evolving autonomous system might do. The action filter architecture proposed in this chapter, which provides a hard "not to exceed" boundary to range of action, delivers an out-of-band method to audit and edit autonomous behavior, while still keeping it within parameters of acceptability.

References

[1] Vinson (2013). X-47B Makes First Arrested Landing at Sea, Navy.mil.

[2] Casteli et. al. (2016). Autonomous Navigation for Low-Altitude UAVs in Urban Areas, arxiv.org.

[3] Ansari & Alam (2011). Hybrid Genetic Algorithm Fuzzy Rule Based Guidance and Control for Launch Vehicle, Intelligent Systems Design and Applications (ISDA) Conference.

[4] Wang et. al. (2016). Learning to Reinforcement Learn, arxiv.org.

[5] HistoryNet (2006). Lawrence Sperry: Autopilot Inventor and Aviation Innovator, HistoryNet.

[6] Chicago Tribune (1947). Reveal 'Robot' C-54 Zig-Zagged Way To England, Chicago Tribune Sept. 24, 1947.

[7] Welch & Bishop (2001). An Introduction to the Kalman Filter, SIGGRAPH 2001.

[8] Bin & Justice (2009). The Design of an Unmanned Aerial Vehicle Based on the ArduPilot, Indian Journal of Science & Technology, April 2009.

[9] Ferrer (2016). The Blockchain: A New Framework for Robotic Swarm Systems, arxiv.org.

[10] Hexmoor et. al. (2005). Swarm Control in Unmanned Aerial Vehicles, ICAI 2005.

[11] Federal Aviation Administration. Flight Operation Branch, Category I/II/III ILS Information, www.faa.gov.

[12] Kennedy & Eberhart (1995). Particle Swarm Optimization, Proceedings of IEEE International Conference on Neural Networks. IV. pp. 1942–1948.

[13] Karaboga (2005). An Idea Based On Honey Bee Swarm for Numerical Optimization, Technical Report-TR06, Erciyes University, Engineering Faculty, Computer Engineering Department 2005.

[14] Pan & Yang (2009). A Survey on Transfer Learning, IEEE Transactions on Knowledge and Data Engineering.

[15] Luotsinen (2004). Autonomous Environmental Mapping In Multiagent UAV Systems, Masters Thesis, University of Central Florida.

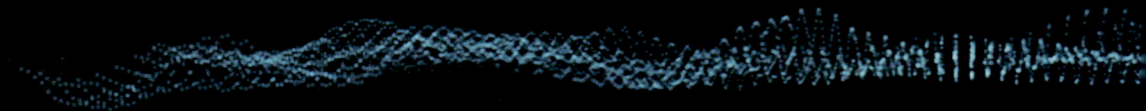

::: CCRDN_H

12.903122, 100.8890

12.896653, 100.8787

13.553979, 100.5365

13.575338, 101.6022

13.061319, 99.80353

25.706166, 102.7433

14.257045, 100.3843

11.858878, 102.9618

12.923593, 100.5368

::: 3D_FORM_VIS

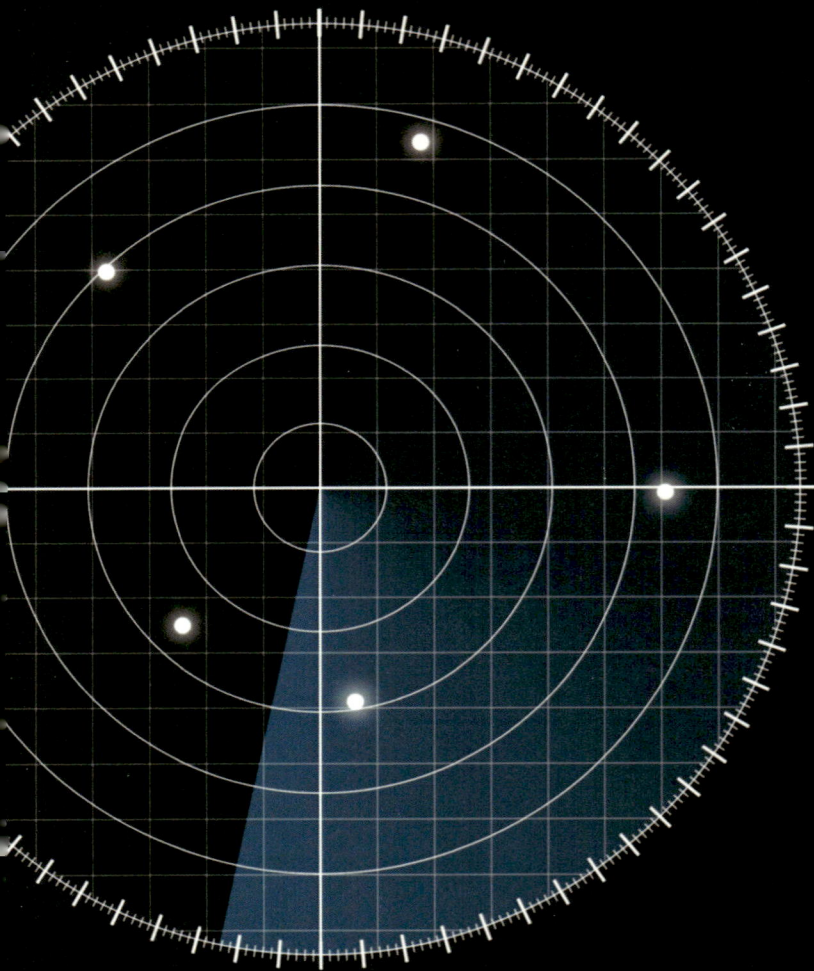

8.
AI and Large-Scale Information Analysis

Abstract

In this chapter, we explore a proposed system that can integrate a large number of diverse streams of information, interpret and categorize them automatically, extract key insights, and present them to a human analyst to form a common intelligence picture (CIP). The automation of a large number of tasks, including interpretation and inference, should reduce human workload significantly, while simultaneously allowing a larger amount of data to be processed and analyzed. The goal is to enable systems that can process "infinite" (or very large-scale, constantly growing) streams and repositories of data. In this chapter, we propose DarkForest, an artificial intelligence-based, multi-source analysis and reasoning system that aims to meet this goal. DarkForest is unique in 1) its use of natural language processing (NLP) techniques for insight extraction 2) the combination of automatically constructed predictive models with NLP and numeric (time series) feeds 3) the placement of information that has been ascribed semantics due to automated processing in layers of context 4) a sophisticated inference engine that derives new conclusions from facts already known to the system and 5) constant monitoring and autonomous updates to the information processed via the overall framework.

Introduction

We propose an AI-based analytics system that can leverage open source, closed source, and multi-format information streams for analysis and insight extraction. This system, codenamed Dark-Forest, is based on AI techniques designed, tested, and field-deployed by SparkCognition, one of the country's leading artificial intelligence companies. The techniques applied include natural language processing, ontology-based reasoning, deep learning, automated model building using a fusion of genetic algorithms and artificial neural network (ANN)-based approaches, graph analysis, heuristic-driven search, and additional techniques.

The DarkForest system and its constituent algorithms is designed to leverage open source software (OSS) systems such as Spark[1], Hadoop[2], HDFS[3], Cassandra[4], Hbase[5], and other tried and tested big data systems. Additionally, custom algorithms are designed to use clusters and graphics processing units (GPUs) for acceleration and large-scale processing whenever possible.

The goal of this chapter is to describe the architecture of the DarkForest system and explain, via a few working examples, the viability of developing and scaling such a system to provide intelligence support to organizations involved in the pursuit and assurance of national security.

Basic System Architecture

Information Streams

The system is architected around the notion of inbound information streams that may be event-based (i.e., which update and pass data through the system when a new piece of data is placed on the stream) or poll-based (lbid. which respond to queries from the processing system and then return the requested data). In the Dark-Forest system, multiple streams of information can be processed

---

[1] Streaming analytics platform
[2] Map-Reduce and cluster management
[3] Hadoop Distributed File System
[4] NoSQL database
[5] Database based on Hadoop

at the same time, and each can include data of multiple types; for example text, audio, video, and images.

Source Adapters

Each stream of information is ingested into the processing system by means of a source adapter. This is a software component that can dynamically be added to enable additional types of sources and streams to be supported. This design also allows for flexibility, error isolation at ingest time, and scalability, as source adapters can be hosted on multiple physical systems.

Big Data Stores

Source adapters provide data that is ultimately stored in big data stores. In the DarkForest system there are multiple types of databases and stores that store information of different types. For most sources, data is placed on a distributed file system (DFS) with minimal processing to minimize latency. Data ingestion processes run out-of-band with other processes responsible for writing data to the DFS. These ingest processes can then place data in relational, NoSQL, key-value pair, graphical, and other custom data stores (e.g., serialized data structures on distributed file systems).

In addition to hosting ingested information, these stores also contain configuration and other dynamic customization elements (including some code stored in stores) that are useful to the proper functioning and configuration of the DarkForest system.

Stream Processing

As mentioned in the Introduction section, DarkForest is designed to leverage open source systems as much as possible to reduce deployment costs and leverage the rapid evolution of such frameworks due to the presence of large development communities around most of these systems. In order to enable processing of data as it enters the system, DarkForest supports stream processing, or the near real-time processing and incorporation of information as it is placed on the "wire" and ingested via an event-based adapter.

In order to support stream processing, DarkForest makes use of the Spark open source analytics engine.

Semantic Interpretation

The raw data that is brought into DarkForest for processing has to be interpreted for it to have meaning for analysts, and for automated reasoning processes that are implemented as multiple AI algorithms, within the system. For example, a piece of text data that might be a news story contains the names of certain key officials, cities or weapon systems. Such a story may even contain multiple such places, actors and entities. In order for the text to be overlaid with more specific meaning, i.e., the fact that a certain set of words is actually the name of an individual, semantic interpretation is necessary. We explain below a few of the techniques that are used to do this within the DarkForest system.

Ontologies

Ontologies are essentially hierarchies of semantic entities – categories or sets/subsets – that can be used to organize information concerning a domain or subject. As an example, an ontology representing the architecture of government in the United States may start with the idea of government, which then is connected to three child nodes representing the judicial, legislative and executive branches of government. Further nodes would then show the entire hierarchy of governmental

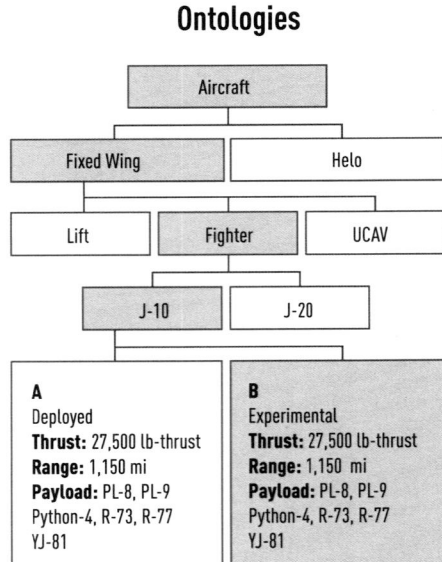

Ontologies

Aircraft

Fixed Wing | Helo

Lift | Fighter | UCAV

J-10 | J-20

A	B
Deployed	Experimental
Thrust: 27,500 lb-thrust	**Thrust:** 27,500 lb-thrust
Range: 1,150 mi	**Range:** 1,150 mi
Payload: PL-8, PL-9	**Payload:** PL-8, PL-9
Python-4, R-73, R-77	Python-4, R-73, R-77
YJ-81	YJ-81

organization. Now, this is a very specific structure, and since not every country follows the United States' model, it may make sense to generalize such structures so that they fit with a larger number of possible data sets.

Generalized ontologies can be useful because they can establish relationships between and among entities, they can identify properties or attributes that are expected of elements (nodes) within the ontology, and they also provide a general framework for reasoning processes to implement actions and rules without being concerned with any specific instance. For example, a reasoning process that involves the notion of "head of state" needn't worry that the head of state in Saudi Arabia is a monarch and in the U.S. this role is taken up by an elected official called "president." The reasoning processes may also not need to concern themselves with the specific names of these individuals, but instead represent all the additional information about the abstract concept of "head of state" as attributes of a specific instance of a generalized node in an ontology.

Frames

Frames and slot-and-filler structures in general are a technique that has been used in AI in many different contexts, successfully. A frame can represent a scenario, the expected actors that are relevant to the scenario, the expected relationships between these actors, and additional context if necessary. For example, the frame of a cabinet meeting will entail the presence of members of the cabinet and the head of state. The expected outcome of such a meeting is deliberation or some decision set. When mention is made in a news story to a cabinet meeting, all this additional context may not be captured in the story itself because it is assumed that a human reader would understand and recognize the background and nuance.

However, in the context of an automated system, this is not a good assumption. The presence of certain terms, or more sophisticated deep learning-based word vector embedding techniques can be used to process a piece of text and determine which frames are "triggered" or relevant for the text in question. By selecting frames, the DarkForest system gains additional context and background that is not directly referenced in the processed text at all. This can allow questions to be answered with greater precision and clarity, and relationships to be established where no direct relationships would have been uncovered were the input information to be taken at face value.

Model Building

Automated Extraction of Ontologies and Construction of Frames

One of the criticisms of the use of ontologies and frames is that the process of putting these together takes a lot of time. The famous AI project, Cyc, has been active for the past 30 plus years and has not achieved its original goals of encoding "common sense" as a set of logical rules. In large part this is the case because the process of building such a system is manual, labor-intensive, and not particularly scalable.

We aim to utilize automated techniques for the extraction of terms of art, relevant actors, and objects and relationships between them in order to build ontologies in an automated fashion. Of course, ontologies constructed in this automated fashion will be subject to human approval. Inputs received from human "approvers" and analysts will be factored back into the system to make it more accurate in producing future ontologies.

Predictions

SparkCognition has developed and deployed numerous systems that utilize sophisticated algorithms for predictive purposes. The range and nature of events predicted is diverse – from the expected failure of an industrial asset, on the one hand, to the expected future value of a bond or financial instrument on the other. A broad range of predictive algorithms will be incorporated into the DarkForest system, including time series prediction methods that leverage deep artificial neural network structures.

Classification

Particularly in the context of text analysis, the SparkCognition team has developed numerous algorithms that allow the processing, categorization, and classification of text samples into a set of defined or undefined categories.

DeepNLP

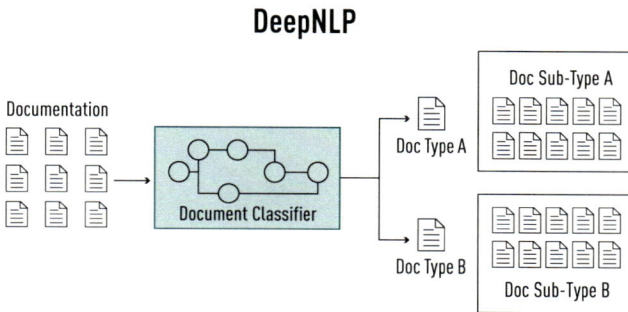

In the event the categories have been defined (e.g., an ontology of document types or topics), the system is able to ascribe new documents to these pre-defined structures. However, in the event no such ontology exists, the documents are processed and placed in machine-named topic sets. Human users can then label these topics and either go with machine-ascribed relationships between them, or define new relationships.

Visualization and Interaction

Information that has been processed and classified has context within the DarkForest information model. This context specifies whether the information pertains to individuals, places, or objects. The visualization of these information types is handled with a multi-layer model that allows different types of entities to be isolated, selectively combined with others, or viewed together with all other known information as a single unified whole.

Large-Scale Visualization

In order to support the visualization goals described in the previous section, working in concert with data that is massive and multi-dimensional, a very scalable visualization system is necessary. DarkForest will use WebGL and OpenGL (Open Graphics Language)-based, highly-scalable rendering systems to show the information necessary. Additionally, even with the use of these libraries and APIs it is expected that some scenarios will require yet more optimization. The SparkCognition team has had experience with additional optimization and scaling techniques and has developed solutions to some of the expected scale challenges as it pertains to visualization.

Deep Q&A

Deep question and answer (Deep Q&A) is a capability that allows human users to simply ask a question in natural language and receive in response a precise and specific answer from the system. This answer is usually a few sentences which come closest – given the information the system knows – to answering the question asked. This approach is contrasted with a typical search engine that matches keywords and provides a list of links, or document references, for the human user to peruse, analyze, and read. The answer is ultimately found by the human user – and for a large-scale document set where a large number of questions must be asked before an insight or outcome is discovered, the search engine methodology simply doesn't scale.

The IBM Watson system is one AI-based platform that aimed to provide sophisticated deep Q&A capabilities. SparkCognition has developed a proprietary system called DeepNLP that can extend systems like Watson and can also provide completely stand-alone services that are uniquely differentiated. DarkForest will contain integration points with DeepNLP, allowing it to support answers to natural language questions.

Event-Driven Proposals

It is not always the case that a relevant insight or finding is discovered in response to a human analyst asking a question. An autonomous information exploitation system should be able to interpret new data as it arrives and make sense of it, proposing to human users elements to which attention needs to be paid. These event-driven proposals are another form of insight presentation supported by DarkForest. As new information enters the system via adapters, the information is automatically processed without waiting for any human input. If trends and relationships are discovered that the system knows are of relevance (for example, these

DarkForest

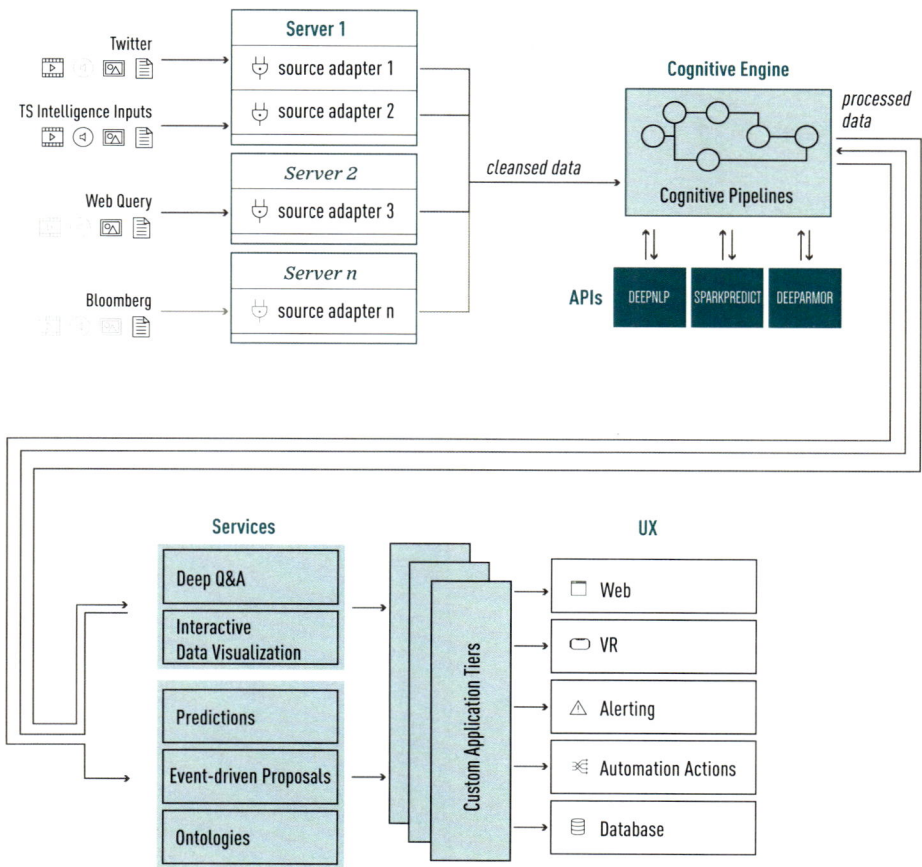

could have been defined by humans directly, or may have been trends/relationships that humans have paid close attention to in the past and the system has learned that fact), then the information is presented or pushed to ensure that it is immediately noted.

Data Source Examples

To power a system like DarkForest and to enable it to make meaningful discoveries, information and data inputs are of key concern. The greater the number and variety of data sources, the better. Following is a description of a few types of data sources we envisage the system supporting.

Internet-Based Data Sources

DNS Records

Domain Name Service records contain information about registered internet domains (e.g., whitehouse.gov), which registrars the domains are held by, when the domain was registered, who registered the domain, and information about the administrative technical and billing contacts. Many times the information for the actual owner of the domain is held private by a domain registrar in lieu of an extra privacy fee. However, much information is extractable from the DNS system despite these privacy options.

Open U.S. Government Records

The U.S. government provides a large number of datasets that can be relevant for many applications. For example, the Dept. of Commerce published a list of banned individuals and entities that are sanctioned and with whom U.S. persons should not conduct business. The presence of the names of such entities in news articles, describing their latest moves or dealings, may be of interest. The Dept. of Commerce dataset is just one, but there are literally thousands more, ranging from county tax and property records to trade data and litigation records.

Social Media Streams

Social media streams are often useful in that they bring an immediacy that can be unmatched. For example, the first mention of the Bin Laden raid did not come from a news station or Reuters, but from an individual Twitter user who saw helicopters flying overhead and reported the activity immediately. More recently, the Iranian missile attack against ISIS was reported not by the Iranian government, ISIS officials, or a news agency, but by ordinary citizens who started posting videos of the nighttime ballistic missile launch on social media.

SparkCognition has extensive experience integrating social media streams of a variety of types both in terms of the social networks that have been tapped, as well as the types of streams extracted. Sometimes, in addition to the passive listening to data, it is possible to pose questions and obtain responses on these networks, adding very specific answers to aid in analysis.

News and Analysis Websites

An obvious source of relatively reliable content is newsfeeds from established news organizations. These include organizations such as Reuters, AFP, and Bloomberg, and can also extend to private analysis firms such as Stratfor.

Dark Web/Deep Web

For data of certain types, the dark and deep web can be a rich source of information. For example, after many doxxing attacks hackers post proprietary or confidential information on dark web locations and servers. Monitoring these locations can provide access to information that is not immediately known to others, and the trends in the growth/posting of such information can itself be valuable.

As an example, the Guccifer doxxings, DNC hacks, numerous individual corporate hacks, etc., are all usually posted to dark/deep web locations.

Forums, Chat and Discussion Groups

A large number of special interest forums run by nationals of almost every country provide an avenue for enthusiasts to post their views and learn more about topics of interest. However, it is at times surprising how the participants of these forums can sometimes be in possession of privileged or otherwise unknown information. Often, this information is disclosed online. Examples that we have witnessed are pictures of protests the government in a certain country would otherwise seek to suppress, or military movements – pictures of military equipment as it is being carried from place to place, and even pictures of new military systems not previously seen. Many Chinese weapon systems have become known to western analysts through such (deliberate or unintentional) leaks, as an example.

Images and Online Videos

As mentioned earlier, text reports from most sources are augmented now with images and online videos uploaded to free services such as YouTube. A sophisticated system with access to large-scale information processing capabilities could, for example, match faces and build databases of government functionaries, protesters, and other actors of interest. Gait analysis on video could also be useful when applied to videos that have been posted by terrorist or insurgent groups.

GIS and Satellite Information

A vast amount of satellite data is available now through three principal means. First, free and open sources that might not be very high resolution and may not be refreshed frequently, such as Google Earth. These sources are useful nonetheless as they can serve as a "free," easy to obtain layer that can anchor other types of information visualized with DarkForest. The second source of data is commercial, but unclassified information obtained from imagery providers such as Digital Globe. This usually has higher resolution and is refreshed more frequently. The third source of data would be from intelligence agencies such as the NGA (National Geospatial Intelligence Agency) that would provide very high fidelity, near real-time data.

Regardless of the source of satellite map data, DarkForest incorporates such information into both the visualization subsystem, enabling its display as a configurable layer, as well as in the internal models and reasoning system to anchor other data that has geo-location attributes – for example, the discovery that a large number of pro-Daesh social media messages originated from a particular location moments before, or after, a major attack was unleashed. This type of automated correlation and discovery allows accounts or persons of interest to be identified and analyzed more deeply. It also allows certain locations to be associated with phenomena or trends of interest.

RF and SIGINT

One of the information layers we envisage the DarkForest system will support is an EF or SIGINT (Signals Intelligence) layer. Depending on the application, this could include the location of RF emitters such as Wi-Fi access points, or radio transmitters and stations, or recorded sources of RF-initiated IED activations.

Unconventional Sources

Unattributable SIGINT

Unconventional platforms such as inexpensive weather balloons integrated with small embedded microcomputers and requisite sensors (cameras/CCDs, directional antennas and radio equipment, etc.) can be operated in a potentially unattributable way. The onboard software sufficiently obfuscated using anti-tamper methodologies, a small platform flying at high altitude could potentially overfly a protected airspace undetected. In the event that it is shot down or otherwise fails over surveilled territory, there would be nothing of consequence or markings of any importance other than equipment that would look like hobbyist gear. Additionally, altitude-based self-destruct (both software and hardware) can also be potentially implemented to further reduce the chance of the system being used as evidence.

It is worth noting that the highest reported altitude a simple weather balloon has ascended to is nearly 180,000 feet, which is a significant altitude far above the ceiling of fighter aircraft and even high-altitude surveillance aircraft. While some SAM systems, such as the SM-6, designed for the interception of ballistic missiles, can bring down a craft operating at altitudes in this range, the cost of such an intercept would be tremendously high, and if successful, hardly anything would remain of the craft.

Distributed Sensors

Distributed sensorlet systems that are integrated with a Common Intelligence Picture system such as DarkForest can provide inputs that include vibration, audio, visual, and other types of sensors. A sensorlet system such as SparkCognition's "perceptive dust" concept demonstrator can be used to create a peer to peer, robust, low-cost network distributed over an area of coverage to collect signals intelligence. For example, vibration sensors on board such devices can be used to detect IED implantation events, or the audio sensors on board can detect a particular type of vehicle going past or a drone flight overhead.

These sensors can provide feedback into the DarkForest system, which can leverage algorithms such as those implemented in SparkPredict, to use distributed sensor information to identify activities and events via machine learning algorithms.

Analysis Examples/Use Cases

Identifying Research of Interest

Consider an example where DarkForest is employed to identify practitioners and researchers related to a key area of investigation in a nation of interest. Performed manually, this can be a laborious task. Just the initial research necessary to make a list of such practitioners would involve reading through thousands of publications. Further, the list of such publications would grow so quickly that any repository would quickly become dated unless it were continually refreshed at great expense. It would be ideal to:

- Automate the sourcing of candidate research material for analysis

- Automate the sourcing of news items and other reference materials (including social media) to determine names of researchers and practitioners

- Autonomously analyze documents produced by these researchers, or ones that cite their work, to continuously develop a graph of relationships between work produced, the researchers responsible for putting this work together, the institutions and agencies responsible for supporting such work, key figures in those agencies, and even collaborating scientists involved with the principal researchers in the work of interest

- Constantly update the lists and repositories by continually scanning source materials

A system could be used to maintain well-structured databases of lists of persons of interest based on criteria concerning subjects/research areas deemed sensitive. Confidence levels regarding the accuracy of each entry in the list, and a fully "explainable" trail of how each record made its way into the database, can also be included to enable data quality while providing assurance and accountability.

Other Examples: Early Warning

There are numerous similar applications that involve automated research that can be easily constructed as pipelines within the DarkForest system. Sometimes the information collected from open sources can be incredibly timely and unique. Note that the first indication that the Osama Bin Laden raid was in progress occurred when a Twitter user in Abbottabad, Sohaib Athar, transmitted the following message:

"Helicopter hovering above Abbottabad at 1AM (is a rare event)"

The Bin Laden raid is not the last time privileged information will make its way out into the open via a public, electronic channel like Twitter.

In fact, the missile strikes carried out by Iran against ISIS were filmed and logged via Twitter and Facebook. This was done almost instantly, and included video indicating the number of missiles launched and the location of the Twitter user. Such information can be crucial both for intelligence gathering purposes as well as for early warning in the event that an information breach appears to have occurred, or an underway mission appears to have been compromised.

A DarkForest-based "early warning" capability would involve large-scale live data analysis coupled with Natural Language Processing to identify posts, tweets, and stories of interest. The capability can also be expanded to include non-public information sources.

Knowledge Graphs

As described in the preceding sections, information is analyzed as it enters the DarkForest system and entities of interest are extracted automatically. However, it is not just the entities that are extracted but also their relationships. There are two principal sources of relationship extraction. First, there are frames which define the analyst's view of how certain types of entities fit together. For example, a president has a cabinet, and the cabinet has a minister of state, and the president can appoint said minister. In this context, both the entity relationships and an action one entity can take with respect to another is encoded in a frame.

The secondary source of relationships is derived from the automated analysis of the data itself. A variety of machine learning and statistical techniques are employed here. This includes co-reference, concurrent mentions, statements that when analyzed with natural language techniques indicate relationships (e.g., "Barack Obama's daughter's name is Malia Obama") and term relationships extracted via word vector analysis (a deep learning technique that uses word vector embeddings to determine the relationships of terms to each other). This type of word vector analysis is particularly useful because it has reasonable success in analogizing and, given that Trump is the president of the United States, can learn the answers to questions such as, "Who is to Iran as Trump is to the United States?" Given access to raw information that contains mention of the Head of State of Iran, this deep learning-based technique can analogize and extract entity relationship information that was not directly fed to it.

Based on both these broad categories of relationship sources, we propose to overlay a knowledge graph on top of the raw data, almost as a higher-order index that can be independently viewed, traversed, linked to and reasoned over.

Graph-based reasoning, however, is not the only form of reasoning the DarkForest system would support. Additional context is provided in the following section.

Reasoning

We propose to embed a Reasoning Engine at the heart of the DarkForest framework. This engine allows the derivation of new insights from raw data by following rules of logic, as well as statistical and machine learning principles. A multi-step process would enable additional insights to be generated from incoming data, per the following flow of activities:

- Ingest data

- Process data and store into data stores

- Out-of-band, analyze stored data to extract semantics (entities etc.)

- Create entity relationships and represent as graphs

- Out-of-band, analyze data to identify predicates (Donald Trump is the president of the United States)

- Create confidence levels for extracted predicates (Predicate 1 is confirmed by all sources that mention it, Predicate 2 has been seen only once, Predicate 3 has majority agreement by sources, Predicate 4 has minority agreement by sources)

- Use predicate confidence levels to filter subset that is used to derive additional facts

- Produce second-order and higher-order predicates and add them back to the predicate data stores

- Use ML and deep learning techniques such as word vector embedding to analyze sentiment, tone, and topic fits for inbound data

- Create ontologies and display topic visualizations to identify shifting tone and topic of analyzed discussions

- Automatically identify correlations in time series and NLP trends, finding waveforms that fit well together or appear to have robust correlations that may suggest causation

While the above is not an exhaustive listing of the types of reasoning and analysis processes the DarkForest system will perform, the outline should serve as an indication of the breadth of reasoning techniques we plan to employ.

Conclusion

The design for DarkForest integrates a multi-algorithm, pipeline-based analysis approach for knowledge applications. The system is designed for handling multiple types of information and reasoning across these diverse sources. Further, the system can infer new facts or potential insights from the ingested information. DarkForest also allows for conversion of language data into numeric forms, such as time series, to allow correlations and potential causations to be analyzed. A powerful, multi-layer visualization system enables the identification of complex or related phenomena with clarity and ease. Finally, the combination of SparkCognition's proprietary AI algorithms for analysis and highly scalable open source infrastructure for execution provides the best of all worlds in terms of sophistication of analysis, low cost, and proven volume handling.

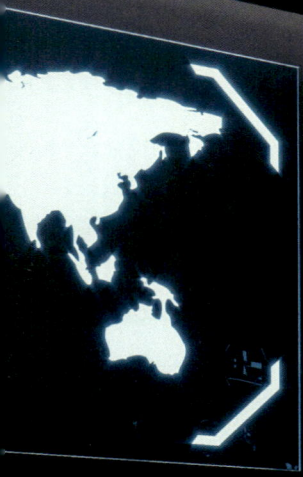

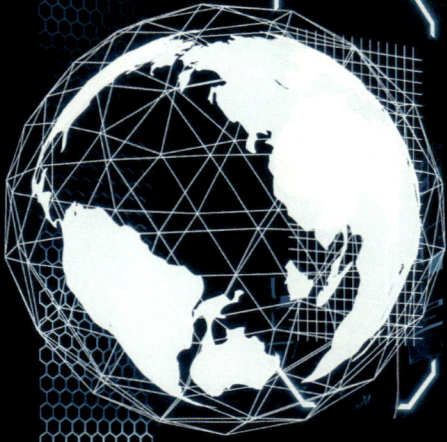

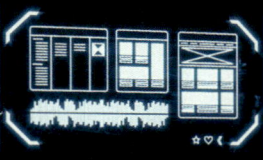

9.
AI and Quantum Computing: An Interview with Amir Husain

What difference are AI and quantum computing making to security, warfare, and the conduct of military operations?

Artificial intelligence has the potential to revolutionize warfare at all levels: from the micro decisions made by a smart projectile as it recognizes targets, avoids counterfire, and eviscerates its objectives, to the highest level of automated analysis and fusion of ISR flows that inform senior commanders. Gen. Allen and I have written extensively on this subject, and our "On Hyperwar" piece (chapter four of this book) published by the U.S. Naval Institute *Proceedings* journal outlines in detail our projection of AI-driven impacts to the future conduct of military operations. In a nutshell, AI brings distributed, federated autonomy to the battlefield at a scale unparalleled in history. It enables the coordination of massive numbers of individual systems with far greater confidence and accuracy, delivering a concurrency that has never been seen in war. It tightens the OODA loop to such a degree that a smaller force leveraging AI can perpetually operate inside the enemy's decision-making cycle, delivering a David vs. Goliath effect at speed and scale before the enemy has had a chance to register or respond.

Quantum computing is at a relatively early stage of development. Large companies like Google and IBM are making investments, alongside startups such as Dwave, but a viable quantum computer that is useful for a variety of large-scale tasks is not presently possible. In general, when this technology matures, quantum computing has the potential to undermine cryptographic constructs that have long secured military communications. QC also has the potential to create such massive speedups in computational activities that require parallelism that many of the supercomputers in use today will seem like calculators for the many tasks that quantum computers will perform well. One area of active research is the implementation of AI algorithms and machine learning techniques on simulated quantum computers, in anticipation of the real developments in quantum computing hardware allowing these algorithms to run at quantum computer speeds. In other words, work on the software has started.

Another area that leverages quantum effects is quantum communication, and here, the practical developments seem further ahead. China has deployed the first quantum communications satellite which provides the ability to conduct tamper-free communications... if any third party inspects the message en-route, the intended receiver will know immediately. This is a capability traditional communication systems have lacked.

Is NATO lagging behind and to whom? To the U.S.? To China and Russia? What are these "others" doing? If NATO does lag behind other military powers is it a case of "preparing for the wars of the past instead of the wars of the future?"

NATO does appear to be lagging behind. Probably to China, Russia and the U.S. China is developing AI-powered missiles, has deployed quantum communications satellites, is investing in an increasing amount of machine learning technology for military uses such as sensor interpretation, and has just published a national policy document that commits billions to ultimately position China

106

as the top AI power by the year 2030. Europe and NATO have no joint position on this subject, much less programs in place. The EU funded one large AI project some years ago, Henry Markram's BRAIN project, which was research into reverse engineering the human brain. Much more is needed, at multiple "harvest" time horizons... you need things now, in five years, in 10 years and further out than that. All together. The U.A.E. just announced a new ministry and appointed the world's first Minister for AI.

Russia, which has always had excellent programming talent and cybersecurity technology (e.g., Kaspersky is one of the largest security companies globally), is investing in autonomous UGVs (Unmanned Ground Vehicles) developed by the Kalashnikov bureau. Dmitry Rogozin recently shared a video on his social media profile showing a short demonstration of a humanoid robot firing handguns at a target. But the real threat from Russia in the near term is their very successful weaponization of information. Combined with their offensive cybersecurity capabilities, this has the potential for disrupting democratic institutions and national partnerships and undermining trust at a global scale. Perhaps we saw this play out in the last U.S. election. The leverage of AI and its sub-field of natural language processing technology – well within Russian capability and skill sets – will automate the threat, providing far greater scale and impact.

The U.S. is the birthplace of artificial intelligence, home to a large number of research universities that run AI and robotics programs (greater in number than the European Union) and has, until recently, been the primary source of origin for AI start-ups. China is attempting to best the U.S. in many of these areas, but most NATO member states – seen independently – would be distant competitors on these metrics. NATO minus the U.S. would still not be competitive. The U.S. is pursuing many areas of AI research through its research and rapid acquisition offices, such as DARPA, SCO (Strategic Capabilities Office), DIUx (Defense Innovation Unit Experimental) and service specific RCOs (Rapid

Capabilities Offices). These areas extend into important fields such as xAI (Explainable AI) which seeks to explain many of the "black box" decisions that emanate from successful connectionist systems such as the popular "deep learning" methodology.

In my understanding, NATO lacks a rapid acquisition office and suffers from a five year (at minimum) procurement cycle. As I mentioned to the NATO officials in Brussels, and in my remarks at the GLOBSEC/NATO meetings, this will be the kiss of death when dealing with a technology like AI which – in some areas – is doubling in capability every few months.

NATO has (finally) taken seriously cybersecurity and cyberspace, to the point of acknowledging cyberspace as the fourth military domain. Are AI and quantum computing a further development of cybersecurity – or a qualitative leap?

AI is not just about cybersecurity. It is about the automation of decision-making and the automation of action at post-human speed and scale. It is thus applicable to all domains. Its ability to process information flows more massive than any human can comprehend, to remember everything ever seen and to train itself over time – sometimes even by competing with itself using techniques such as generative adversarial networks (GANs) – has no similar parallel. It is the automation of MIND. I have often spoken publicly about the fact that the human race basically needs to do just two important things. First, we replicated MUSCLE, with the steam engine. And that achievement led us out of a basically miserable past that showed miniscule progress over vast periods of time into the industrial age. We are now at the cusp of the second goal: replicating MIND. And we are doing this already, at least in part. When we get to certain level with AI, the resulting cognitive revolution has potential beyond imagination. Will it happen in three years or 73? It almost doesn't matter at the timescale most countries should concern themselves with. In the near term, narrow AI – or the automation of specialized cognitive function in specific domains – is already outperforming humans. Some of these areas, such as piloting aircraft in difficult situations, are incredibly relevant to the conduct of military operations.

Quantum computing provides a new type of computer – a new type of massively faster hardware – on which many types of computational processes and algorithms can run. As I noted earlier, many machine learning algorithms will be adapted to this infrastructure as it becomes available. The resulting speed and capability increases will lead to what we call a double exponential... one exponential trend riding another exponential trend. In mathematical terms, you are dealing with something that can tend to infinity pretty quickly.

Can non-state actors develop an AI capability? In other words, can AI and quantum computing become yet another asymmetric security threat? The experience of the last 20 years shows that unlike WMDs (nuclear, biological, chemical weapons) new weapons like IEDs have become easily available to terrorists and have become a powerful military equalizer between State and non-State actors. Can the same happen in the area of AI?

Yes. The Houthis, ISIS, and the Taliban are all known to be using commercial drones. They've used them for surveillance, as flying IEDs and for propaganda. Today on the internet, there are open fora for autonomous control software for many types of commercial drones. Even relatively simple (everything is relative, of course) adaptations of commercially available hardware and known algorithms (e.g., computer vision algorithms), integrated on consumer drones can create a significant capability. Something that is within the realm of the possible TODAY is a terrorist group using a large number of inexpensive commercial drones to launch a relatively large-scale, coordinated attack relying on autonomous piloting instead of individual human controllers. The net result will be massive kinetic impact created by a relatively small number of people. And perhaps even by an individual.

10.

Putin Says Russia's New Weapons Can't Be Beat. With AI and Robotics, They Can

Russia's next-generation of strategic weaponry may be a bit more distant and a bit less fearsome than Vladimir Putin recently claimed. But his March 1, 2018 speech about titanic ballistic missiles and nuclear-powered undersea drones should spur American defense and technology communities to move faster – indeed, uncomfortably so – to embrace similarly disruptive ideas such as artificial intelligence and robotics.

America's adversaries are betting that a new wave of weapons will negate technologies and tactics at the heart of U.S. military might, among them aircraft carriers and high-altitude missile defense. Russia's newest weapons, Putin claimed, are "invincible against all existing and prospective missile defence and counter-air defence systems." China's defense investments follow a similar path, with an aggressive testing tempo for hypersonic weapons, unmanned aircraft, and advanced submarine detection, among other capabilities. Even if Putin's coming arsenal doesn't quite live up to its hype, the U.S. should nevertheless understand that America's adversaries will soon field weapons like the ones he described – perhaps even before the Pentagon does.

There are, of course, conventional ways to respond to such threats, rooted in over 70 years of Western defense engineering and domestic and alliance politics. Yet these new Russian weapons are intended less to pulverize than to provoke. They are meant to draw a response that will further reinforce Putin's narrative of an encircled nation threatened by NATO and U.S. missile defense systems.

To avoid this trap, then, the United States ought to seek unconventional responses. Some promising concepts are made possible by recent advances at the intersection of artificial intelligence and robotics.

One particularly dramatic moment during Putin's speech featured an animation of a 200-ton Russian Sarmat ballistic missile releasing multiple warheads toward targets in Florida, purportedly enough to obliterate a region the size of France. The traditional approach to stopping such a country-killer is with a sea- or ground-based ballistic-missile defense system, the sort that are being deployed within the U.S. and in allied nations like Japan and Poland. But these defenses require feats of technical marvel to work correctly. Moreover, deploying them to allied bases can present domestic challenges, as in South Korea, and provoke Moscow or Beijing.

But now imagine that the apocalyptic Russian video of reentry vehicles streaming toward Miami, Orlando, Tampa, and Palm Beach takes a different turn. The camera zooms to an unmanned submarine surfacing 10 miles to the east of Miami. Within moments of breaching the light blue Atlantic waters, the vessel's clamshell deck doors spring open. Another dozen submarines positioned off the southern Florida coast emerge in similar fashion, and together they launch hundreds of quadcopter UAVs.

The electric-powered drones dash skyward, initially cued to the incoming Sarmat by thousands of U.S. Air Force wafersats. Within moments, each group has created an encrypted local network that replaces easy-to-jam GPS navigation signals with optical and self-referential inputs. A few moments more, and the swarms refine their plan of attack. One group over Tampa splits to join up with the swarm forming over Orlando, guided by a data packet from an F-35 out of Eglin Air Force Base.

This insight is crucial. The F-35 reports that the reentry vehicles are hypersonic Avangard models, now streaking earthward at Mach 20. The swarms quickly array themselves in aerial layers, deepening the defense. Drones carrying fragmentation explosives switch places with the thermobaric-armed UAVs, which sprint to an even higher altitude to meet the incoming warheads. Tens of thousands of feet below, the unmanned submarines are back underwater, their onboard neural networks reconfiguring the weapons payloads and form factors of the next wave of UAVs to better respond to more Sarmat missiles or another aerial threat.

This kind of land- or sea-based defensive mothership-swarm operational concept could be used against other aerial threats, such as the long-endurance nuclear-powered cruise missile Putin revealed, or underwater against the Status-6, a roving nuclear-armed torpedo-like drone designed to evade traditional anti-submarine defenses.

There are technical challenges, to be sure, such as power management or deploying a cheap and resilient global sensing network. But they are not insurmountable, nor is this kind of countermeasure hypothetical. In fact, SparkCognition began working on this very swarm-mothership concept a few years ago and has filed U.S. patents covering the design of such systems. Moreover, the same advances in machine learning algorithms that make drone-launching robot submarines a reality can also create global data-gathering networks based on sensors that cost less than last year's mobile phone.

AI and robotics – the very forces that are ushering in the era of "Hyperwar," as one of the authors and General John R. Allen, USMC (Ret.) call it – already allow U.S. asymmetric responses that are inexpensive, resilient and globally scalable. Ultimately, though, the biggest challenges with autonomy and robotics will not be technological. It will be our willingness to break with convention.

─────────

This chapter is based on an article that was originally published in *Defense One*.

11.
Security Challenges Posed by Improvised UCAVs: Contours of a Counter Strategy

Abstract

In this chapter, we outline (a) plausible scenarios whereby enemy improvised unmanned combat aerial vehicles can be employed against civilian and military targets, (b) current counter-UAV technologies on the market and their limitations, and (c) a proposed system to more effectively neutralize improvised unmanned combat aerial vehicle (IUCAV) threats.

Introduction

Both military and law enforcement contexts is the neutralization of IUCAVs. As has been demonstrated in conflicts in the Middle East and by hobbyists everywhere, it is quite possible to rapidly modify commercially available unmanned drones to carry small payloads, perform surveillance, carry out real-time direction of fire for artillery and missile forces, and even act as aerial snipers.

Before we consider ways to counter IUCAVs in urban and open environments, let us first discuss some plausible scenarios that have either already occurred or can easily occur in the future. By "easily" here, we mean that the technological means to realize these scenarios is open source and the authors (at least) are familiar with means to achieve these aims without the need for any classified or prohibited technology.

Following our analysis of use cases, we will discuss the methods employed by current systems designed to foil UAVs. And lastly, we will propose several mechanisms we feel are essential to mount a credible defense in the age of large-scale, reasonably intelligent, autonomous drones.

Scenarios

Scenario I: Urban Swarm Assault in a Densely Packed Environment

For this scenario we consider a combination of concurrently piloted UCAVs, which may be flying under autonomous guidance, or controlled by humans at a distance. These weaponized commercial drones are flown into a large, densely populated structure such as a stadium. The attackers use multiple vectors of ingress and attack exit points first, impeding any emergency exodus. Improvised anti-personnel explosive payloads could be employed to cause maximum damage.

In the seconds following this initial onslaught, a round of additional fixed wing IUCAVs carrying heavier payloads begin to dive in at high-speed following a set of waypoints to approach the venue and then finally tear into the target once it is in range. Given the large size of a stadium and the very confined space of the rows of seats, precision isn't essential for these craft to have a terrifying and crippling effect.

Since fixed wing craft can cover greater distances and carry greater payloads, they make it possible for the attack to be launched from 10 or 20 miles away, increasing the area that must be searched by law enforcement to track down the perpetrators.

In the third wave of this attack scenario, fixed wing UCAVs descend on the parking lots that usually surround a stadium-like public venue. Some UCAVs autonomously identify and attack densely packed crowds of fleeing spectators, while other UCAVs identify exposed vehicles and attack into their rear intending to explode the fuel tank. Alternatively, quadcopters with explosive charges could also trigger charges in close proximity to the cars designed to ignite the fuel in these tanks. In a large public lot where hundreds or thousands of cars might be parked, this form of attack would create massive panic, and masses of blast and burn casualties complicating the work of first responders and creating a mass casualty crisis for the neighboring hospitals. The effect and the optics of this attack would be traumatic for the local population, but also the nation.

The systems used to execute such an attack would require half a dozen UAV pilots, or alternatively, drones equipped with autonomous control, a few dozen IUCAVs, and commonly available Google satellite map data, as well as GPS data gathered by walking and scoping the target area prior to the site of attack. The explosive materials employed would likely not be difficult to source given the accessibility of such materials to past lone wolf attackers and the presence of instructions on the internet.

Scenario II: UAV Attack on Large Commercial Airport and Aircraft Infrastructure

Much like in scenario I, a combination of quadcopter and fixed wing UAVs that are either remotely piloted via FPV (first person view) equipment, or routed via GPS waypoint guidance, can be used to launch an assault against commercial civilian airport infrastructure. A tactical advantage can even be gained by employing these IUCAVs at night against a flight line of aircraft waiting to take off at a major airport. In this situation, the aircraft usually move very slowly and are hemmed in with planes ahead and behind. Doors are locked and passengers secured with seat belts.

Again, if the airliners' fuel tanks can be targeted (in the case of commercial airlines, the wings), the kinetic effects and damage might be multiplied even further and communicated across multiple waiting aircraft.

An additional disruption in a scenario such as this might be for fixed wing UAVs to also attack the control tower. IUCAVs with explosive charges rigged to blow on nose impact could fly directly into the glass panes of a control tower at high speed. The effect of this attack would be to close this airport, creating a ripple effect across the entirety of civil aviation, and leading to the airport's evacuation, creating masses of densely packed and panicked passengers and airport employees. (See scenario I.)

Once again, an attack of this type could be carried out with the perpetrators miles away from the actual airport. The present security mechanisms found at most civilian airports would be useless in defending against this type of threat.

Scenario III: Explosive UAV Piggyback Attack Against Commercial Airliners

In this scenario, a UAV rigged with explosives approaches a taxiing aircraft at night and attaches itself to the underbelly or wing of the airliner via a tether or vacuum coupling. As the aircraft gains altitude, the perpetrator can detonate the charge. Or the charge can be automatically triggered at a certain speed or altitude.

Scenario IV: UAV Swarm "Bird Strike"

A simpler attack than scenario III, a swarm of UAVs can be directed into the path of an airliner while it is taking off or landing. Any UAV strike on the external surface of the aircraft, or a UAV being ingested into an engine, if not immediately catastrophic, would create an immediate in-flight emergency and an emergency divert to the closest field or return to the airport. As with Scenario II this attack would likely close this airport until authorities were able to sound an "all clear" subsequent to their investigation. The effect of the attack and the disruption would be dramatic. Any aircraft that survived the attack would be down for maintenance for a very long time, if not written off.

Scenario V: UAV Flying into Buildings for Targeted Killings or Event (Crowd) Targeting

In this scenario, fixed wing or quadcopter systems are flown into a modern skyscraper building with glass windows to target specific individuals who might reside in offices or rooms close to these windows. While many skyscrapers use strengthened panes which may not be easily breached by the impact from a single commercial quadcopter, an IUCAV can be adapted to carry a tactical glass shattering tool. If this tool is used to impact the glass with force, it could create a breach through which an incendiary or explosive payload could be delivered either by that IUCAV or by a follow-on system.

Scenario VI: Assault on a Major Urban Highway When Clogged in Rush Hour

Assets and strategies similar to scenario I and II can be employed against highway structures such as multi-level flyovers which, when clogged, offer no escape. The effect of the attack would be multiplied by drivers fleeing en masse on foot, or in the likely panic of drivers ramming cars to attempt escape, or even driving off such flyovers into traffic below.

Scenario VII: Assault on Cellular or Radio Infrastructure, or other Communications Infrastructure

In combination with any of the techniques identified previously, small commercial UAVs can be used to disable cellular and wireless infrastructure in a limited area by damaging highly visible tower-mounted antennas. Additionally, we can envisage emergency radio communication jamming techniques that could be employed in parallel. As an example of how this might work, machine learning algorithms can be fed signals from an SDR (Software Defined Radio) to detect emergency transmissions on the radio spectrum. Once the band(s) being used for such communication have been identified, a transmitter could be automatically configured to hop to these emergency bands and begin jamming. While the effects of this type of jamming will be felt in a limited area, cutting out cellular and emergency radio communications while another kinetic attack is underway can complicate the situation and amplify impact. (See scenarios I and II.)

Scenario VIII: Assault on Power Distribution Infrastructure

Today, utility companies use commercial quadcopters to remove unwanted debris and entangled obstructions from power cables and grid infrastructure. An attacker could employ simple, specially designed munitions deployable via IUCAVs that attach themselves to power cables and use small quantities of thermite to melt live cables. In a more sophisticated version of this attack, autonomous IUCAVs can be used – empowered with vision algorithms – to deploy a large number of these charges along a long run of commercial grid infrastructure, causing public safety hazards, as well as multi-hour power breakdowns.

Scenario IX: Assault on Gas, Chemical, and Fuel Storage Facilities or Transport Vehicles

Similar to scenarios I and II, IUCAVs can be used to land on above-surface fuel storage tanks or large fuel transport vehicles and detonate a charge designed to ignite the fuel contained therein. What would differentiate this type of strike from a terrorist launching an RPG against a similar target is the relative difficulty of obtaining an RPG in a domestic context, compared to the relative ease of obtaining a commercial drone and the necessary chemicals. Secondly, the range at which the effect could be achieved is far more than with an RPG. Third, the RPG is a LoS (line of sight) direct fire weapon, limiting the vantage points a terrorist would use to attack a target. In comparison, the IUCAV can be flown via FPV (first person view) optical link technology in small numbers, or via autonomous GPS coordinate-based guidance in large numbers. Fourth, IUCAV attacks against such facilities can be structured so that a large number of these craft impact multiple targets in a relatively short period of time, or possibly, concurrently. To achieve a similar effect, a large number of attackers would need to locate themselves in viable direct fire positions without detection.

Scenario X: Assault on High-Value Targets (HVT) in Transit

Small commercially available quadcopter UAVs have speeds exceeding 70 MPH and can accelerate rapidly, some even as fast as 0-60 MPH in 1.1 seconds. A sufficiently large frame to carry an explosive payload can be used in quantity to attack a high value target such as a convoy of vehicles, or HVT personnel as they are slowly traversing an open, public area such as a street. Beyond the impact of the first IUCAV, attackers could clone the successful path/trajectory and program it into multiple small IUCAVs to continuously descend on the target. The impact would be that of a dozen hand grenades suddenly appearing from seemingly nowhere, descending at high speed from above and exploding with no warning.

Beyond Specific Scenarios: Generalized Threats to Consider

While the scenarios listed above are intended to tangibly exemplify the threats posed by commercial-off-the-shelf (COTS) UAV technology, it is not possible to list every individual scenario in a study such as this. Future threats may borrow from some elements of the scenarios above and integrate unforeseen approaches, or remix several of the ideas presented here to materialize a unique threat. It is thus useful to consider some of the technologies and tactics that can be used beyond these specific scenarios. Our hope is that the discussion in the next section will allow readers to imagine their own list of possible or likely scenarios and strategies to guard against them.

Multi-Vehicle Execution of Dynamic Learned Paths

Using FPV (first person view) technology, it is possible for practiced pilots of high-speed hobby drones to weave in and out through dense foliage, avoid obstacles, and perform impressive aerobatics. When an attacker employs such sophisticated piloting capability for an IUCAV, the craft will be able to race through dense urban environments and follow unpredictable paths that make it impossible for defenders to easily compute a firing solution to counter these threats.

This is an ominous outcome, but many may look for hope in the fact that attackers likely won't find it possible to have too many such expertly piloted UAVs operating simultaneously. Unfortunately, we see no reason to be so hopeful. A capability we call "multi-vehicle execution of dynamic learned paths" is a potential force multiplier for an attacker and would address this shortcoming, making future IUCAV attacks far more lethal.

The basic idea behind this approach is for a human pilot to guide a first IUCAV and for the system to learn behaviors in real time, allowing other IUCAVs to autonomously follow the first craft to impact the same target area with unpredictability, speed, and precision. As the controls are initially manipulated by the human pilot, a machine learning system watches and learns from control movements, intervening waypoints, sensor information regarding UAV status, and GPS/inertial location. The ML (machine learning) system can capture the observed "pilot workflow," save it, and adapt it rapidly for other available IUCAVs so that they impact the target in rapid succession.

As a consequence, once a single IUCAV has been manually piloted – creatively and aggressively – to a target, the human controller moves on to a new unique piloting task while the saved workflow is repeatedly adapted and applied to other IUCAVs, causing them to be autonomously guided toward the target in question. The net-result is that a single expert human pilot can have an outsized impact on multiple targets, conducting a veritable campaign with a single pair of hands on a joystick.

Note that with the employment of this approach, full AI-enabled autonomous control of the IUCAV isn't required. As a result, this approach could manifest in a real-world environment quickly.

Computer Vision and Autonomy

Today's powerful yet power efficient RISC-based (reduced instruction set computing) processor architectures and miniaturized, mobile-device-ready GPUs (graphics processing unit) provide lots of opportunities to integrate high-end machine learning and computer vision capabilities aboard compact, off-the-shelf hobbyist UAVs. Vision capabilities can be used to identify targets, find assets of interest as a part of autonomous search missions, perform vehicle identification and monitoring, and deliver budget ISR capabilities to an asymmetric foe.

Cyber and EW Payloads for Commercial Drones

The authors' own past work has focused on developing autonomous cyberattack systems that can automatically identify targets on a network and launch attacks designed to penetrate and disrupt systems and networks. Integrating such automated hacking technologies with UAVs means that Wi-Fi networks high on the 50th story of a building not otherwise accessible to an attacker can now be attacked. Many commercial and power utility systems that contain integrated Wi-Fi capabilities can be scanned and attacked via UAV-borne automated hacking software. Wi-fi systems can be jammed too, or false SSIDs (Service Set Identifier) can be advertised at strategic locations via IUCAV-borne access points that capture the credentials and traffic of users who naively believe they are connecting to a known access point.

Night Operations

Commercial night-vision cameras and IR equipment can enable IUCAVs to conduct autonomous night operations, when they are difficult to spot and stop. Any safety measures built into commercial UAVs, such as flashing ID lights, can be easily disabled by an attacker to take advantage of the small size, relative silence, and night capability of modified commercial UAVs.

Laser Guidance for Small UAVs

Using inexpensive, commercially available lasers it is possible to develop beam-riding control systems for hobbyist-level UAVs. UAVs modified in this way can be used as projectiles, easily guided with the simplest of control requirements, i.e., the aiming of a beam onto a target. In addition to the ease with which drones can be controlled with this mechanism, multiple IUCAVs can ride this beam simultaneously or in rapid succession, allowing attackers to multiply their control bandwidth and saturate the target or quickly switch from one target to another.

Staged, Scheduled Execution

Orchestration of a complex attack with multiple physical attackers can be difficult, particularly when the action is spread over a large area. Unless attackers are incredibly well practiced and precise, it can be expected that some element of their timing will be off, allowing own forces to find an opening for counter action. However, this is another area where, with the use of algorithmic scheduling and precise location sensors, distributed, large-scale coordination across a large number of vehicles becomes possible. As a result, the defender's task increases in complexity.

Shielding

While many counter-UAV technologies rely on RF jamming, we believe that RF shields and autonomous control software will make such technologies ineffective. The concepts of shielding and making UAVs immune to RF jamming is discussed in more depth later in this chapter.

Current Anti-Drone Technology

Anti-Drone Rifles Using RF

Kalashnikov in Russia, Battelle in the U.S. and TAI in Turkey have all developed anti-drone rifles that emit a radio signal directed at an unauthorized, approaching drone. The core idea behind these devices is that jamming of the control signal will cause the drone to land. Indeed, landing when control is lost is the default programmed behavior in most hobbyist and commercial UAV systems.

However, there are numerous shortcomings to this approach. First, the anti-drone rifle is an intensely manual and slow weapon that must track the drone all the way until it lands. Second, it is not likely to work at all against a determined attacker who has either shielded the drone against RF jamming or has made the guidance autonomous, and thus not reliant on remote control. A particularly insidious programmer could actually home in on a jamming signal and race towards the source to neutralize the jammer.

Shotguns

Shotguns can be effective at close range, but they are manually operated and thus suffer from limitations in locating and tracking of potential targets. They are not likely to be successful against a large number of inbound objects and they are also easily foiled by drones taking randomized final approach trajectories as they close in on their targets. Some companies have developed cartridges that can be fired from standard shotguns and deploy anti-drone nets. These have questionable efficacy in tests and are viable only at very short ranges.

IR SAMs, MANPADs Such as Stinger Missiles

In battlefield scenarios such as the Yemen war, the Stinger or Patriot SAM could present options to neutralize drones. But this is a trade any enemy will make all day long; the missile rounds are individually far more expensive than 100 improvised hobbyist drones! An additional issue with these systems is that they have limited simultaneous engagement capability and can thus be overwhelmed and quickly saturated when confronted with a swarm. Finally, they are not practically employable in a safe manner in domestic urban and law enforcement contexts.

Railgun

The Navy's in-development magnetic railgun will indeed enable the targeting of drones, but like SAM batteries, railguns are likely not deployable in commercial or domestic urban environments. And while they do extend the range of engagement, it will likely not be possible to equip a single platform with more than a single railgun. Therefore, swarms will always be able to overwhelm, or optimize approach from areas outside the arc of fire. Multiple railgun platforms (fleet) may mitigate these challenges, but then the cost and complexity of defense skyrockets in the face of an inexpensive, guerrilla-style threat.

Lasers

Lasers are a viable multi-shot system that can target drones – effectively, melting them. However, laser systems are expensive, they require specialized power, and they need to stay focused on the target for some time in order to disable it. This practically translates into a constrained rate of fire and leads to overall limitations regarding the number of such systems deployable at scale.

Trained Eagles and Falcons

Some groups have attempted to train birds to identify and take down drones. The less said about this approach the better. If this approach were even remotely viable, nefarious drone operators would equip drones so that they would cause great harm to a difficult-to-train bird upon contact. The bird would perhaps take down one drone but would then itself be lost as a result. Drones are far more easily replaced than these specially trained animals!

Net Guns and Shooters

Some companies have developed man-portable systems that shoot nets to trap drones. These are effective at very short ranges. Adaptations include carrying such net shooters on larger drones which then fly toward the smaller drone. Once again, a single shot system like this is fundamentally not scalable, of limited effectiveness due to its short range, and easily countered by even a modest-sized swarm.

Shortcomings

The collective shortcomings seen across current counter-UAV approaches are summarized as follows:

- Expensive
- Difficult or unsafe to employ in urban environments
- Lack bandwidth to take on a volume of targets
- Slow per-target neutralization sequence
- Impractical due to slow replenishment of counter projectile
- Short range
- Centralized detection
- Manual dependence
- Inappropriate assumptions regarding effects of jamming

Countering UAV Threats with an AI-Powered Integrated Defense System

We propose an area denial system for IUCAVs and small low-speed craft that is based on five system elements:

- Integration of data from existing radars
- Network of fixed and mobile distributed multi-sensor systems
- Data fusion and airspace management system integrated with machine learning-based object identification
- Autonomous UCAVs that can support completely automated CAP (combat air patrol) missions to develop a cordon sanitaire around a key target at a key time (e.g., football game)
- An urban-safe, autonomous CIWS utilizing pneumatic (non-chemical) means of launching massive numbers of light pellets

Data From Existing Radars

Some large commercial UAVs can be picked up by today's radar network. Accessing this already available information to integrate into a comprehensive full-spectrum picture would be the first logical step in developing an early warning capability.

Network of Fixed and Mobile Distributed Multi-Sensor Systems

The existing radar network, while useful, does not provide full coverage at all levels and was certainly not designed to guard against threats such as commercial UAVs. Therefore, we propose to augment it with a network of distributed sensors, including optical, RF, and acoustic sensors. These can be integrated with dispersed infrastructure such as cell towers, utility poles, traffic cameras/light posts, and private/public buildings. These distributed sensors would not perform any location computation. They would simply capture and relay information to a central location, where machine learning algorithms can be applied to make sense of whether a signal corresponds to an inbound UAV.

Data Fusion and Airspace Management System Integrated with ML-Based Object ID

As detailed in elements one and two above, the information gathered from existing radar networks as well as from our proposed network of inexpensive, distributed multi-sensor systems can all be integrated to form a single coherent operating picture. To form this picture, extensive use of advanced machine learning will be required, where the approaches will not be limited to mere vision and signal-processing algorithms, but will extend into reasoning and complex multi-sensor input-based predictions. Confidence levels and underlying evidence (explainable AI insights) will further refine the picture and provide data necessary to plan and prioritize counter actions.

Autonomous Counter-Air UCAVs

One of the most compelling and unique pieces of the counter-IUCAV system we envisage is the employment of our own specially adapted IUCAVs in a counter role. The autonomous control system we propose for small UAVs would enable these systems to fly CAP missions to detect inbound, unauthorized craft, relay their positions, and even engage them in a defensive counter-air role.

Some of the authors' existing work is directed toward the development of precisely such an autonomy capability.

An Urban-Safe, Autonomous CIWS Utilizing Non-Chemical Means of Projectile Launch

We also propose an urban-safe ground-based defensive system which will take the form of a pneumatically-powered, computer-controlled close-in weapon capable of firing lightweight projectiles the size of small caliber ammunition or "BBs." The authors have developed a basic design for a high-pressure system that can fire thousands of pellets a minute. These lightweight pellets would be designed to have minimal impact when they land in urban settings. Marrying such a pneumatic cannon with a computer-controlled alt-azimuth mount, target data feeds, multiple optical sensors, and computer vision algorithms will yield a low-cost, yet urban-safe, effective defensive system that can be employed against inbound IUCAVs that manage to slip through own counter-air IUCAVs.

Conclusion

We are about to see a proliferation of commercial UAV and related technologies, such as embedded sensors, machine learning algorithms, and low-cost, compact, and low-power hardware optimized for running these algorithms. Employment of these systems with some creativity will enable attackers to launch deadly strikes, such as those envisaged in the scenarios covered in this chapter. Existing anti-IUCAV defenses are weak for multiple reasons, including their cost, their manual operation, and their limited capability in dealing with swarm threats. An effective, multi-tier defensive system that can provide comprehensive protection against such future threats in both urban and battlefield contexts is absolutely critical. We believe that a distributed network of optical, RF, and acoustic sensors connected to machine learning algorithms optimized for target detection can provide an effective early warning capability. This early warning capability can be used in concert with an urban-safe, high rate of fire and yet inexpensive anti-IU-CAV CIWS as well as an autonomous, small form-factor defensive counter-air drone to enable multi-tier, full-spectrum protection. What we present in this chapter is not limited to concepts alone. We are well on our way to developing many of the capabilities discussed herein.

12.

As AI Begins to Reshape Defense, Here's How Europe Can Keep Up

Change comes hard in much of Europe, particularly in the defense community. But no less than in the United States, European nations are wrestling with the implications of machine learning and artificial intelligence – in the military as well as civilian society. During several trips to Europe in the last six months, we have noted a significant uptick in the number of NATO political and military leaders discussing AI's impact on the alliance's military capability.

There seems to be a two-speed discussion going on. European defense industry officials we talked to had no qualms about harnessing AI to reduce manufacturing costs and improve customer satisfaction. But governments and institutions like NATO and the EU were having a harder time. Will AI's impact on society – say, in data privacy – be feared and, hence, regulated? Can it be "purchased" for national defense or domestic use, and how much would this cost a tight-fisted government? Could it, perhaps, simply be ignored?

One problem we observed during our trips is that "AI" means different things to different people in Europe, just as it does in the U.S. During one of many dinners with journalists and business leaders in Brussels, it was variously described to us by the attendees as "hoovering up personal data" from across Europe or as the "secret sauce" or as "magic dust." And while all agreed that harnessing it was not a simple matter of "buying three boxes of AI," there was little consensus on how governments and institutions could or should integrate this new technology.

Not all NATO member governments are spinning their wheels. In Paris, the French government under Emmanuel Macron has announced it will spend $1.85 billion over five years on AI-related research and development, including the creation of a French DARPA-like agency. This effort is not just related to the military field; President Macron became energized about the technology after visiting healthcare facilities, where AI-infused data analyses was improving patient care. He also senses that French companies may be able to harness AI for, say, autonomous driving. In an interview with Wired magazine he said, "I think these two sectors, I would say, healthcare and mobility, really struck me as promising. It's impossible when you are looking at these companies, not to say, 'Wow, something is changing drastically and what you thought was for the next decade, is in fact now.' There is a huge acceleration."

Across the English Channel, the importance of AI was recognized early on by Britain's business community and its government. With a one billion-pound public-private AI-industry investment initiative launched last month, the commitment is real. The UK defense establishment regards its use as yet another pressure on their strained budgets – but perhaps also a data-analysis tool to reduce operating and maintenance costs.

Germany has a different problem. As in Britain, France, and the United States, the German commercial sector is hotly pursuing AI as a means to offering cutting-edge products at lower prices; but compared to its NATO allies, this commercial R&D finds its way less frequently to the government and defense sectors. Public concerns that AI might give the German military or intelligence services too much power or insight into private lives have made Angela Merkel's coalition reluctant to jump in with both feet.

NATO and the EU can play an important role in helping Allies understand and absorb the role of AI. Both institutions can also be consumers. The follow-on aircraft to the aging NATO AWACS fleet will doubtless involve AI. But at this stage in Europe, the most important role EU and NATO can play will be as advocates and validators.

First, however, these institutions must see clearly for themselves the role for AI – not just in military arsenals and counterterrorism toolboxes, but in other sectors of government as well. Unlike NATO, the EU will play a major role in regulating AI applications like self-driving cars. This undertaking will likely last years as the EU tries to keep pace with change. Regulating AI applications may be a comfort to some governments whose politicians fear change, but too much regulation will stifle development and Europe will be denied some of the advances (and business) AI will bring about. There will have to be a balance reached by governments on both sides of the Atlantic between regulation and freedom, a journey that will take years.

So what more can the nations of Europe do to take part in the AI revolution? How can NATO allies avoid widening the interoperability gap with the U.S. military? These six elements are central:

Don't Wait

Change is hard, and it is easy to avoid difficult decisions by waiting to see what governments are doing. AI itself is a topic that could launch a thousand meetings. But there isn't time for the 100 percent solution. AI is here now and advances come in such leaps that spending time on one aspect will cause another to speed by unnoticed. Gen. John R. Allen, USMC (Ret.) and AI thought leader Amir Husain have noted that while the U.S. both birthed and incubated AI technologies, the Chinese and Russians are not far behind. European governments need to tackle AI now and be part of the conversation and not wait to see what other governments will do.

AI is Not "Buy America"

AI is not a U.S.-only technology or a marketing technique to sell U.S. products, but a global phenomenon. In addition to China and Russia, others are also capitalizing on advances in AI and machine learning. Dismissing AI as an American-only initiative is unsupported by the facts and is a sure way to fall behind. There are signs of progress, at least, such as last month's European Commission appeal to bump up AI-related investment to at least 20 billion euros by 2020.

AI Needs an All-of-Government Approach

AI and machine learning are not just for the military. There are many parts of government that can use the technology to make better decisions, save money, and provide better security. For an all-of-government approach to work, stovepipes between sectors must be broken so that R&D can be shared and applied across governmental sectors.

No Fortress Europe

It will be tempting to protect a nascent European AI industry by locking out U.S. or other foreign tech companies. But this would be self-defeating; the technology is growing so fast that partnering – on application as well as development – is the only way to stay ahead of the curve.

EU and NATO are Critical

These organizations must guide their member nations, gathering and disseminating best practices on applying AI. NATO defense planning can ensure that allies take advantage of new military capabilities, such as new-generation drones or intelligence analysis, or new operating efficiencies made possible by AI data-crunching. The EU can also nurture AI by avoiding onerous regulations before AI is even applied. Finally, either or both groups might launch their own DARPA, or AI center of excellence.

Shape the Rules

Yes, there are healthy debates about the more esoteric visions for AI, such as autonomous machines and computers. But that is no reason for any country, or for institutions like NATO or the EU, to avoid AI. Nations must be in the game to help set the rules to avoid the concerns that others may have. Nations cannot sit this out and then expect to have their concerns heard. In fact, at the end of the day, their – and our – national security depends on it.

———

This chapter is based on an article that was originally published in *Defense One*.

13.

Emerging 'Hyperwar' Signals 'AI-Fueled, Machine-Waged' Future of Conflict

Imagine wars fought by swarms of unmanned, autonomous weapons across land, air, sea, space and cyber. The autonomous weapons use artificial intelligence-based algorithms to make decisions, advanced sensors to maneuver and pinpoint precise vulnerabilities in targets and offensive and defensive cyber capabilities – all in real-time and independent of human decision-making.

This is the emerging nature of warfare, and it's portrayed via a series of vivid vignettes in a recent article in the U.S. Naval Institute's *Proceedings* Magazine article, "On Hyperwar," authored by U.S. Marine Corps Gen. John R. Allen (ret.) and Amir Husain, CEO of AI company SparkCognition and author of the book, The Sentient Machine. Fifth Domain recently caught up with Husain, who gave an in-depth interview on this concept and its implications for near-future warfare.

Generals and military theorists have sought to characterize the nature of war for millennia, and for long periods of time, warfare doesn't dramatically change. But, occasionally, new methods for conducting war cause a fundamental reconsideration of its very nature and implications. Allen and Husain, in "On Hyperwar," identify cavalry, the rifled musket and Blitzkrieg as three historical examples.

For Allen and Husain, the advent of Hyperwar signals the next "fundamentally transformative change" in warfare. The term Hyperwar has been used historically with different denotations, which Allen and Husain trace, but the authors have redefined and adopted the term to describe this "AI-fueled, machine-waged conflict." Allen and Husain wrote:

"What makes this new form of warfare unique is the unparalleled speed enabled by automating decision-making and the concurrency of action that become possible by leveraging artificial intelligence and machine cognition. ... In military terms, Hyperwar may be redefined as a type of conflict where human decision-making is almost entirely absent from the observe-orient-decide-act (OODA) loop. As a consequence, the time associated with an OODA cycle will be reduced to near-instantaneous responses. The implications of these developments are many and game changing."

"On Hyperwar" details some of these changes, which include the concepts of infinite, distributed command & control capacity, concurrency of action/perfect coordination, logistical simplification and instant mission adaptations.

In his interview with Fifth Domain, Husain said he has been interested in AI since a young age. More recently, he has been intrigued by applications of autonomy in weapons systems. Running an AI company, Husain is at the forefront of applying new AI-driven technologies. For instance, he compared the company's SparkPredict AI-based machine failure prediction system to building R2D2 from Star Wars. He compared the company's Deep-NLP (natural language processing) system to building C3PO from Star Wars.

"What was science fiction a few years ago is becoming fact now," Husain told Fifth Domain.

Husain and Allen started discussing the concept of Hyperwar a few years ago. Both men had been thinking independently about the increasing degrees of autonomy in war – Allen from a strategic perspective and Husain from a scientific perspective. Husain had been working for years on defensive and offensive autonomy in cybersecurity when he met Allen. These mutual interests led to one of their earliest discussions on the concept of Hyperwar, and the two have continued collaborating. Below is a full transcript of the Fifth Domain interview with Husain.

Will Hyperwar force us to rethink the concept and strategy of deterrence? If so, how?

In a nutshell, yes, I believe so. I say this because the basic calculations that go into determining a minimal, credible deterrent might well shift. For example, how many enemy ballistic missiles can our BMD (ballistic missile defense) systems be expected to intercept? These hard calculations will change in the era of Hyperwar.

Once opponents incorporate AI into guidance systems and deploy counter-BMD AI, the numbers that constitute a deterrent will change. And that's at the high end. The types of protection we have to afford to strategic systems, such as a Patriot battery, will also change. So far, we've only heard of militias flying remotely piloted commercial drones into radars, a clever strategy designed to temporarily disable a SAM (surface-to-air missile) system. Now, imagine what happens when these drones are autonomously controlled.

Finally, consider the aircraft carrier, the big stick, which has been an important instrument of foreign policy for the United States since World War II. It was impossible for most adversaries to seriously challenge an American carrier strike group. But with AI-powered cruise missiles, autonomously controlled swarms of small, fast moving hydrofoils, an array of UCAVs (unmanned combat aerial vehicles) deployable in swarms and sophisticated information processing algorithms that extend detection and forewarning capabilities, will this continue to be so?

Will Hyperwar change the traditional notions of military superiority/dominance and military advantage? If so, how?

We've long expected that the next near-peer conflict will begin with a cyber salvo. But in the Hyperwar context, these salvos can be machine-initiated and -controlled and thus occur at an expanded scale, increased speed and broader scope.

Hyperwar is essentially about the leverage of AI and machine autonomy to shrink the OODA cycle to nothingness... so tight that it becomes almost impossible to keep humans in the loop in most places. Commanders can continue to supply intent, but the prosecution of much of the war can conceivably shift to machines.

As I mentioned earlier in context of the deterrence question, traditional military superiority and dominance will have to be re-evaluated. A fourth-generation fighter is expected to dominate against a third-generation threat when both are piloted by humans. But when you strip the third-generation fighter of its cockpit, implement sophisticated autonomy and no longer have to pay heed to the physical constraints dictated by the man in the cockpit, outcomes may change.

Consider also that a big source of military advantage is not about equipment, but about training and the quality of your human resources. High levels of training and readiness require immense amounts of capital in the traditional context. The U.S., being the richest country in the world for so long, had a natural edge in this department. But autonomous algorithms don't need that type of time or expense. It is not too long before these algorithms will be evolved in synthetic, simulated environments and then deployed in the real world. When an inferior foe can field highly skilled "pilots" that never tire, don't need training and exhibit none of the biological constraints of a human pilot, what becomes of training-enabled advantage? It shrinks, in my view.

How will predictive analytics influence the start and conclusion – or, as Clausewitz would say, the suspension – of wars, and what role will humans play in those decisions?

Statistical simulation methodologies have long been employed in military planning. With artificial intelligence systems, the granularity at which outcomes can be modeled will become finer. In general, predictive accuracy will increase. AI can be applied to every stage of the war and almost every activity.

Gen. Allen and I spent quite a bit of time coming up with areas where today's AI might drive significant planning and execution advantages. There are dozens upon dozens of opportunities we've identified. For example, prior to the initiation of hostilities, AI can play a major role in how intelligence is processed, and how much raw data – imagery, text, advanced sensor inputs – can be analyzed. We've got the capability to deploy a huge number of sensors, but the way we process this data is bandwidth constrained. We still need humans to do a large amount of grunt work. By moving to more and more sophisticated autonomous information processing systems, we'll increase our capacity to analyze information streams. More accurate and holistic intelligence appraisals are at least one way in which AI systems will influence the war before it's even begun. As for humans, choosing to initiate hostilities is a decision that rests with them.

How will differing "norms" for warfare, perhaps encoded in algorithms, affect thinking on and strategy for military defense?

This is a massively complex topic, but I'll share a bit of my perspective. One of the advantages of employing autonomous systems in war is that the SOPs (standard operating procedures) that optimize warfighter safety can be tweaked to err in favor of reducing accidental casualties.

To clarify your use of the term "norms" of warfare, let's first establish that manned and unmanned systems must comply with the laws of armed conflict. Any entity that fields systems that violate these laws must be prosecuted and punished per international law. That said, in order to minimize collateral damage, autonomous systems can take even greater risks to their own safety than these laws call for.

One application that can potentially make warfare more effective and yet, less deadly, is the use of swarming autonomous systems to seek out a specific target, validate presence to a high degree of confidence and attack with incredible accuracy but limited firepower. Taking out a specific terrorist commander, for example, may only involve eliminating that one individual, not risking innocents in whose vicinity he hides. That level of risk, validation and precision is hard to pull off with manned systems or even with remotely piloted, non-swarm systems. In the future, this may change.

It is common now to see technology advancing faster than human-derived systems can keep up – for example, legal, regulatory, ethical, etc. Which human-derived military systems must be rethought and revamped, or altogether discarded, given the imminent age of Hyperwar?

I believe we'll see increasing levels of autonomy in surveillance and reconnaissance systems as remotely piloted systems begin to morph into autonomous entities. Pulling the trigger may still require a human in the loop for the foreseeable future particularly in low-intensity conflict or during times of overt peace. However, keeping these systems aloft and prosecuting non-kinetic missions will become an increasingly automated process. The information they transmit back will also be interpreted with AI-powered common intelligence picture systems.

While all wars are unique, and each uniquely complex, what will be the single or few critical factor(s) that will likely determine victory in Hyperwar?

- Assimilating and fully understanding the asymmetric impact of autonomy technologies and reflecting this in planning.

- Investing in training, in terms of enabling warfighters to fully leverage the technology of Hyperwar and to develop counters for enemy employment of these systems.

- Finding and optimizing the precise balance of man/machine
  integration. Too little human control, too soon, and we risk compromising transparency and safety. Too much human control,
  and we'll suffer at the hands of tight enemy OODA loops.

*In which technological areas should the U.S. government priori-
tize funding to retain current military superiority?*

Broadly, artificial intelligence. Specifically, machine vision, natural language understanding, explainability, autonomy algorithms,
AI-assisted logistics and planning, automated knowledge management and retention systems and AI-enabled prescriptive/predic-
tive maintenance, to name a few.

Aside from AI, continued sensor development – a key input
to smart algorithms – next-generation propulsion technologies
to enable both efficiency in conventional systems and to enable
practically deployable hypersonic vehicles, space-based systems,
autonomous platforms – aerial, surface and sub-surface – robotics
and investments to enable higher degrees of mobility.

Which traditional military capabilities will Hyperwar make obsolete?

I'd rather not reflect on this in too much depth, but I will highlight a couple of areas. The U.S., thankfully, has the world's
most powerful navy, but there are many countries that employ
non-networked surface ships with minimal air defense capability.
Pre-Hyperwar, these assets wouldn't necessarily be highly surviv-
able, but in the age of swarming, low-cost autonomous drones
and unmanned sub-surface vehicles, you might as well never have
these ships leave port. They could be disabled with minimal expense
and risk.

The employment of offensive cyber systems will rapidly render
useless sensors and air defenses fielded by less sophisticated foes.
The traditional SEAD (suppression of enemy air defenses) mission
and use of stealth jets may in some cases be obviated by a cyber
payload putting a SAM site out of commission without a shot
being fired or a single life being risked.

What are the key takeaways on Hyperwar you would convey directly to policymakers and military leaders?

First, Hyperwar is a consequence of the militarization of artificial intelligence. It is here and will only become more significant with time. We must understand it and factor the implications of broad, widely deployed autonomous systems into our planning and our thinking.

Second, technology has always played a critical role in war, but as we experience a ride on the exponential – nearly asymptotic – technological curve, the rate and extent to which it will impact outcomes in conflict will continue to increase. We must be the best in AI research. We must be the best in the employment of AI systems. The Chinese government recently published their AI plan, with a stated goal to be the dominant AI power by 2030. The dominant AI power won't just dominate the field of AI. Software is eating the world, and AI is eating software... dominance in AI will translate into economic advantage and dominance in the battlefield.

Third, the threats we face in the cyber dimension are about to get more complicated. They will expand beyond information theft, doxxing and cyber physical attacks to mass psyops implemented using autonomous systems. We are about to see an acceleration in cyber capability development, which is being enabled with the application of offensive and defensive AI. This means that we are vulnerable even in times of peace to new and unique threats, which we have not confronted at scale. How many would have guessed that analytics systems and Twitter bots would be used to influence an election and attack the foundation of our democracy? We need to develop a strategy and response framework to deal with the broad spectrum of threats we are likely to see in this new reality.

Clausewitz famously wrote in On War: "The objective nature of war makes it a calculation of probabilities. Now there is only one element still wanting to make it a game. That element is chance.

...But together with chance, the accidental, and along with it, good luck, occupy a great place in war. ...The absolute, the mathematical, nowhere finds any sure basis in the calculations and the art of war. From the outset, there is a play of possibilities, probabilities, good and bad luck... which makes war, of all branches of human activity, the most like a gambling game." Will Hyperwar make Clausewitz's characterization of war obsolete?

No. War – like any large-scale, complex human affair – will certainly continue to be about probabilities. But the purpose of the planner and the commander is to prosecute the war in a manner which minimizes the likelihood of bad outcomes. The technologies of Hyperwar – automated information analysis, broader and deeper intelligence capabilities, low-risk/low-cost autonomous reconnaissance, autonomous cyber systems, higher degrees of precision, to name but a few, will equip the commander with a powerful set of tools to bend and shape the course of the war so as to make positive outcomes more likely.

This piece by Brad D. Williams was originally published in *Foreign Policy* at: https://www.fifthdomain.com/dod/2017/08/07/emerging-hyperwar-signals-ai-fueled-machine-waged-future-of-conflict/

14.

The Next NATO Standard Isn't for Ammunition, It's for AI

When NATO developed Cold War-era standards on everything from ammunition to aircraft grease-ports, commonality of military hardware was essential to ensuring a feasible collective defense of Europe. Today, and into the future, interoperability and joint operations among 29 nations will require a new approach toward the one innovation with perhaps the greatest potential to upend warfare and even NATO itself: artificial intelligence.

AI Everywhere

As an alliance whose potential is defined by the totality of its members, NATO faces a growing challenge in coordinating and collaboration around military-relevant emerging technologies – perhaps even more than during the Cold War when there were fewer member states and a relatively unified threat from the Soviet Union and Warsaw Pact nations. Then, military innovation led defense-relevant breakthroughs, the opposite of today's private-sector driven innovation. AI research and implementation, in particular, is being led by companies, not governments. Moreover, AI will increasingly determine the course of NATO operations, be it in electronic warfare or political campaigning, as Russian manipulation of social media algorithms has shown.

That is just the start, however. Now, and in the future, NATO's missions will be increasingly dynamic and politically thorny – from policing migration to counter-narcotics to strategic deterrence – because of the "algorithmic impact" on political, economic, and social forces around the world caused by machine learning systems commonly described as AI. This can be expected to be amplified by nations such as Russia, which harmonize highly disruptive propaganda and kinetic operations while committing to investments in artificial intelligence. As Russian President Vladimir Putin has said, "Whoever becomes the leader in this sphere will become the ruler of the world." Under President Xi Jinping, China's ambitions are equally grand, and even more credible.

Yet no technology exists in a vacuum. Truly innovative technologies can be simultaneously seamless and disruptive. Nowhere is that truer than for storied multinational organizations like NATO. As this is a larger global challenge, NATO is a perfect "lab" for member nations to find their way forward with AI by embracing commonality through the alliance rather than going it alone.

AI need not add to NATO's recent challenges. Though individual European NATO members have promising government- and commercial-sector AI investment programs underpinned by national strategies as Great Britain and France do, a standardized alliance-driven approach ensures all members benefit from current and future breakthroughs. Unlike defense industrial bases needed for fighters or tanks, critical AI innovations could come from NATO's smallest nations and an alliance-standard approach can ensure these are not crowded out. A set of common NATO AI capabilities matched to the alliance's operating concepts can bridge the technical gaps that could leave out nations without the relevant technology-industry expertise or the ability to implement AI systems at their defense ministries. Moreover, given the complexity of missions around the world, NATO needs to be fully integrated at a mission-systems level, rather than individual states going it alone with incompatible technologies. This is a tenet of the alliance's commonality. Technology development is also moving so fast with AI that any one country's lagging behind in adoption or implementation can undercut whole-of-NATO effectiveness when it is needed most during a crisis. This applies to more than just technology itself: AI's promise is accompanied by growing collective responsibility around data, privacy, and the power of the state. A NATO AI operational framework can ensure world-leading standards are upheld across the alliance.

Indeed, NATO is working on AI commonality, but it is focused on the question of how much freedom to give autonomous machines. "Creating a common standard for describing the role of the human operator and the role of the machine in systems that use AI will help commanders incorporate such systems in their planning processes," NATO officials wrote in setting-up a study due in 2020 on human in the loop considerations for artificial intelligence. "In a coalition environment, such systems potentially deploy in parallel during an operation, which requires that NATO commanders understand the subsequent effect on planning and C2." That is indeed critical, but as important is considering this at a much higher level. With a new "NATO AI Standard," NATO can employ artificial intelligence in a manner that does more good than harm, operationally and bureaucratically speaking.

The NATO AI Standard

A new NATO standard for AI is not a set of measurements or technical specifications, as would be used in establishing, for example, a common caliber for small arms ammunition. This is the era of decentralized innovation and software-driven warfare, when accessible data confers strategic advantage and social media can be as tactically relevant as a light machine gun. This approach to a new NATO AI standard is updated accordingly, akin to a next-generation NATO Standardization Agreement, or a STANAG 2.0.

Rather than being anchored in specific hardware or IT network specifications, the approach is based on the phases of the joint

NATO operational planning framework where each phase has distinct ways that AI can be most effectively used at an alliance level. Additionally, a standardized alliance taxonomy for artificial intelligence needs to be established, based around the understanding that the "narrow AI" systems of today are evolving quickly and will only grow in capability during the coming years.

Indications and Warnings of Crisis

Readiness at an alliance level depends on access to high-quality data to inform intelligence assessments and general situational awareness. It is critical to be able to rapidly share this information in order to develop a common picture that is synchronized at an alliance and national-capital level. One way to accomplish this is to establish AI-focused all-source data processing that is centralized with NATO headquarter elements but produced by, and available to, member states. They have a critical "local" contextual advantage in this analysis, as well as the ability to forward-deploy AI systems. This needs to happen at machine speed, with common operating pictures developed in seconds, not days, to aid overloaded human analysts. Data sources must encompass more than conventional defense-related sources, to include open-source and commercially available imagery, metadata, and social media. Existing machine learning systems already make this possible, but their use is nascent. The current generation of analytical software tools are a step in the right direction, but there is still unprocessed data that is not transformed into national and alliance-level actionable intelligence reports due to a shortage of human analysts. This causes gaps in the intelligence picture, which can be exploited by an adversary.

Crisis Assessment

The transition from pre-crisis to crisis can happen at machine speed. Using AI can smooth out this jarring transition by combining the analytic efforts from NATO command organizations and member nations to ascertain how, and when, a scenario will develop from one phase to the next. Machine learning crisis-simulations systems offer improved visibility into the causes and drivers of a crisis, many of which might be overlooked in traditional military "phase zero" planning or due to insufficient organizational appetite for divergent or highly imaginative models. In a hybrid, multi-domain context, a real-time common operating picture – that draws upon simulations and is also predictive – must extend into the information domain, including modeling peacetime public electronic discourse. The volume and velocity of information during the early phases of a crisis will be nearly overwhelming as the critical factors become clear. Anything other than an AI system has little hope of success in keeping up, while paring away irrelevant information to ensure that human decision-makers are tracking the right information.

Response Options and Mission Planning Development

As with the assessment phase, AI-driven simulations and scenario planning offer substantial insights. With those insights, such systems can create a comprehensive picture of force readiness and logistical imperatives to inform NATO's response options. AI-driven analysis can process massive amounts of unstructured data, maintenance logs, reports, and logistical information to create a detailed and accurate picture of force readiness. The U.S. Air Force is currently using machine learning in a similar manner. Machine learning systems can also fold in data from civilian, commercial sector, and non-governmental sources in order to produce a more accurate "whole of society" capabilities view than is currently possible. Employing this approach among the 29 NATO nations and their respective armed services is especially critical in the context of high-intensity conflict scenarios that are unlikely but potentially catastrophic. Additionally, being able to encompass civilian resources and infrastructure as part of this response option analysis is something that AI systems are suited to doing better than conventional database analysis. In addition to readiness driving response options, another major consideration is developing plans and evaluating their effectiveness. In more tactical scenarios, such as in the development of mission-optimized autopilot capabilities for fighter aircraft, AI techniques such as reinforcement learning are already proven in creating optimized operational plans for a single platform in a similar way that can be applied for entire combat elements.

Mission Execution

The operational execution of a plan by allied forces is based on assumptions regarding the consistency of training, resources, and the joint force's ability to accomplish their mission. AI, particularly when combined with virtual and augmented reality visualization, can play a significant role in providing advanced training and pre-deployment work-ups for NATO-led forces during peacetime to ensure a smooth transition to conducting operations. During those operations, AI "agents" or controllers can be used to assist maintaining, managing, and targeting offensive and defensive systems, ensuring that the joint force is augmented and embedded with autonomous "soldiers" that provide a standardized, and ever-increasing, level of operational competence and consistency of execution. During operations, machine learning systems can use sensor data, entire technical libraries, and advanced models to accurately predict and prevent equipment failure; given the danger improvised explosive devices and precision munitions now pose to supply lines, such efficiency has profound strategic importance. In more tactical scenarios, such as in the development of mission optimized autopilot capabilities for fighter aircraft, AI techniques such as reinforcement learning have shown their utility. Adversarial AI systems running within a simulator can assist in the evolution of a highly optimized, robust mission intelligence that is effective at fulfilling defined objectives. In a similar vein, adaptations of these tactical AI-driven simulator frameworks can be used to gauge likely public reactions to proposed response options, thus allowing planners to leverage AI in order to identify planning loopholes, unexpected consequences, and other unforeseen problems.

Transition

Given the importance of public opinion and political support for NATO as an alliance, as well as its operations, what happens after a crisis is as important as operations during it. Alliance mass communications strategies are reaching peak criticality during disengagement operations, yet even this phase is likely going to be contested. Adversaries already are using AI to do so, to significant effect, which puts the burden back on NATO to defend its narratives with similar technologies. As an example, successful messaging means engagement not just with media but also directly with civil society, from citizens to companies to non-governmental organizations, in multiple countries and in multiple languages. This requires sophisticated and near-real-time interactions on social media, both in terms of pushing a NATO narrative and evaluating the public mood and reactions to outbound communications. Natural language processing systems that can automatically analyze large-scale unstructured data to extract not just sentiment but also model topics, and discover threads and trends for further analysis, are an essential element in future transition activities and establishing a predictive understanding of post-crisis narratives.

Moving Forward, Together

NATO presents an ideal place to develop and establish a new NATO AI standard that has relevance not only among the 29 member nations but as a global benchmark for responsible operations using these new capabilities. Integrating AI into coalition operations is a global challenge for not only NATO but also its supporting allies, many who, in the Hyperwar era, may not even be organized militaries. With the right NATO initiatives in place this diversity can be harnessed to make the alliance more effective and resilient as AI plays a larger and larger role in business and civil society.

Therefore, integrating emerging technologies whose civilian-sector development moves faster than traditional defense procurement deserves renewed attention. One such model could begin with establishing a feeder system that starts to integrate machine learning systems, as well as other highly disruptive but strategic technologies, into member nations as well as the larger alliance commands. It would be integrated with current NATO Science and Technology Organization and standards organizations. As an example, the U.S. Department of Defense's Defense Innovation Unit – Experimental (DIUx) facilitates matchmaking between military customers and commercial-sector companies that develop software-intensive capabilities that can be arduous to acquire with conventional procurement processes.

Machine learning technology will continue to advance in the coming years, if not months, promising breakthroughs that will broaden the capability of AI systems in business, civil society, politics, and conflict. Many advances do not announce themselves, but they can be world-changing all the same. NATO has a perfect opportunity at hand to ensure the alliance is ready for them.

This chapter is based on an article that was originally published in *RUSI Newsbrief*, Volume 38, No. 6. This version has been edited by SparkCognition Press for this book. The Royal United Services Institute was not involved in the production of the current article and is not responsible for its content.

15.

Automating War

The age of artificial intelligence is upon us, with AI systems driving cars, trading stocks, diagnosing illnesses, and performing other life-changing applications. Militaries are also taking artificial intelligence to war, integrating greater autonomy into next-generation weapons. What happens when a Predator drone has as much autonomy as a self-driving car? In the near-future, humanity will have to face the question of whether to delegate life-and-death decisions in war to machines.

Armed robots deciding who to kill might sound like a dystopian nightmare, but autonomy used in the right way could make war more humane. Automation has dramatically improved safety for commercial airliners and is poised to do the same for driving, saving tens of thousands of lives. Could autonomy do the same for war?

Over the past 70 years, automation has already greatly reduced civilian casualties in war. In World War II, unguided bombs had only a 50-50 chance of landing within a 1.25-mile diameter circle, leading nations to blanket entire cities with bombs. Today, precision-guided bombs are accurate to within five feet, enabling strikes directly against military forces and sparing civilians. We've come to expect precision – Human Rights Watch recently argued that firing unguided munitions into cities violates the laws of war.

Future autonomous weapons would go a step further and select their own targets, and they might do it better than humans. AI image-recognition systems have already beaten humans at benchmark tests. Could a robot tell the difference between a person holding a rifle and one holding a rake? The answer is almost certainly yes and, amidst the chaos and fog of war, robots would have other advantages as well. Unlike human soldiers, machines never get angry or seek revenge. They never fatigue or tire. Some have argued that autonomous weapons may someday be so superior to humans at complying with the laws of war that their use would be required.

The precision that is automation's greatest strength is also its weakness, however. Machines will do precisely what they are programmed to do, which can make their behavior brittle. Machines can often outperform humans at specific tasks, but if pushed beyond the bounds of their programming, they can fail badly. Even learning systems that don't rely on rule-driven behavior still only have a narrow form of intelligence, confined to one domain or task. Current AI systems lack the ability to understand the broader context for their actions. Humans, by contrast, are able to apply judgment, ignore rules if they don't apply to the current situation, and flexibly adapt to novel circumstances.

Some decisions in war are straightforward, with an easily identified enemy and a clear shot. Other decisions in war require moral judgments or understanding of the broader context, however.

As an Army Ranger sniper in Afghanistan, I once encountered a small girl scouting our position for Taliban attackers. Under the laws of war, she was participating in hostilities and therefore a lawful target. A robot programmed to obey the laws of war would have killed her. Yet none of the Rangers ever discussed shooting the girl. It might have been legal, but it would have been immoral. No senior commander gave us this rule. Eighteen years of upbringing in American values told us it was wrong. We let her pass and killed the Taliban who came soon after.

Other decisions in war require understanding the broader context for one's actions. Today, Russian and American fighter jets fly in close proximity over Syria, often in tense situations. It would be nearly impossible to program into a machine rules for every possible situation that could arise. Yet human pilots can balance broad and even conflicting guidance, such as the authority to use deadly force to defend friendly forces, if needed, with a desire to avoid starting a war.

The best decision-making is likely to come through "centaur" approaches that combine the speed, precision, and reliability of machines with the flexibility and robustness of human decision-making. One way to do this would be to automate most engagement-related tasks, but keep humans "in the loop" for final approval. There is a risk, however, that an arms race in speed pushes humans "out of the loop" of decision-making, as it has in high-frequency financial trading. Chinese military thinkers have imagined a "battlefield singularity" in which the pace of combat eclipses the speed of human decision-making, leaving militaries no choice but to automate in order to survive. Accidents in stock trading, such as flash crashes, underscore the risks in this approach. In war, there are no regulators to install "circuit breakers" to halt the fighting if it spirals out of control. Nations must put their own safeguards and human circuit breakers in place, or risk warfare evolving into a new domain of automated war beyond human control.

In 2018, countries met for the fifth year in a row at the United Nations to discuss autonomous weapons and what, if anything, should be done about them. Despite calls from prominent scientists and human rights groups for a preemptive treaty banning autonomous weapons, a ban seems unlikely. No major military power or robotics developer supports a ban. Ironically, the rapid pace of progress in AI that alarms supporters of a ban also makes achieving a treaty difficult. Opponents of a ban argue that future machines may eventually overcome their current limitations, and that any prohibition on technology might someday prohibit beneficial uses.

Technology is moving fast, and while progress cannot be stopped, nations need guiding principles for how they should incorporate autonomy into future weapons. This should include an affirmative model for what role humans ought to play in lethal decision-making, beyond simply humans filling in the gaps for what machines cannot yet do. Rather than debate what machines could or could not do someday, we should ask: If we had all of the technology we could imagine, what role should humans play in war? There may be value in handing over control of some targeting tasks to machines, but there are good reasons to keep humans legally and morally responsible for the use of force.

Humans, not machines, are bound by the laws of war, and soldiers have an obligation to ensure that their actions are lawful. This implies some minimum necessary human involvement in lethal decision-making in war. Military personnel must, at a minimum, have sufficient information about the target, weapon, environment, and context for use to ensure that an attack is lawful.

Additionally, there is value in ensuring humans remain morally responsible for killing in war. Arguably, off-loading moral responsibility for killing to machines could lead to less suffering, as soldiers would not have to bear the emotional burden of their actions. But doing so could open the door to far worse consequences. If no human felt responsible for killing, the result could be more killing in war, with more suffering overall, and possibly even more wars.

Recently, Vice Chairman of the Joint Chiefs of Staff General Paul Selva said absolving humans of responsibility for killing is "a fairly bright line that we're not willing to cross." Humans are far from perfect, and there may be ways to use technology to improve on human failings in war. Used in the right way, autonomy could reduce civilian casualties and even warn soldiers if they are about to commit a war crime. At the same time, as nations incorporate greater autonomy into weapons, they will need to ensure that humans remain legally and morally responsible for the use of force. In delegating certain tasks to machines, we must to take care not to delegate our humanity as well. The AI revolution is enabling intelligent machines that can be used for good or evil. With these ever-more powerful technological creations in our hands, it will ultimately be our humanity that provides a check on the worst evils of war.

Humanity stands at the threshold of a technology that could fundamentally change our relationship with war. Machines can do many things, but they cannot tell us what we value, and they cannot answer these questions for us.

Contributors

Amir Husain

Amir Husain is a serial entrepreneur, inventor, and author based in Austin, Texas. He has been named Austin's Top Technology Entrepreneur of the Year, and received the Austin Under 40 Technology and Science Award. Husain is the founder and CEO of Spark-Cognition, an award-winning artificial intelligence company and a member of the Council on Foreign Relations.

Since its founding in April 2013, SparkCognition has received widespread recognition, including the 2017 CNBC Disruptor 50, being named the fastest-growing company in Central Texas by Austin Business Journal in 2017, and ranking on the CB Insights AI 100 list in both 2017 and 2018.

Husain is a prolific inventor with 27 U.S. patents awarded and over 40 pending applications. His work has been featured in outlets such as Foreign Policy, Fox Business News, and *Proceedings* from the U.S. Naval Institute. His book "The Sentient Machine: The Coming Age of Artificial Intelligence" was published in 2017.

Husain served as a founding member of the Board of Advisors for IBM Watson and serves on the Board of Advisors for The University of Texas at Austin Department of Computer Science. He is also a member of the Center for a New American Security Task Force on Artificial Intelligence and National Security.

General John R. Allen, USMC (Ret.)

General John R. Allen, USMC (Ret.) is one of the United States' leading and most seasoned wartime commanders. Gen. Allen served the nation as a Marine Corps four-star General, and as Commander of the International Security Assistance Force (ISAF) and U.S. Forces Afghanistan (USFOR-A). He also has extensive Defense Department policy experience in the Asia-Pacific region, including North Korea denuclearization as well as the relief response to the 2004 tsunami.

His career includes extensive diplomatic experience. In 2014, President Barack Obama appointed Gen. Allen as Special Presidential Envoy for the Global Coalition to Counter ISIL. Before assuming that role, he led the security dialogue for the Israeli-Palestinian peace process as senior advisor to the secretary of defense on Middle East Security. He is a permanent member of the Council on Foreign Relations.

Robert O. Work

Robert O. Work served as the 32nd Deputy Secretary of Defense from 2014 to 2017. He is the Distinguished Senior Fellow for Defense and National Security at the Center for a New American Security, which he led as Chief Executive Officer from 2013 to 2014.

From 2009 to 2013, Mr. Work served as the Undersecretary of the Navy. In this capacity, he was the Deputy and Principal Assistant to the Secretary of the Navy and acted with full authority of the Secretary in the day-to-day management of the Department of the Navy.

In 2008, Mr. Work served on President-elect Barack Obama's Department of Defense Transition Team as leader of the Department of the Navy issues team. He also worked on the defense policy, acquisition, and budget teams.

In 2002, Mr. Work joined the Center for Strategic and Budgetary Assessments (CSBA), first as the Senior Fellow for Maritime Affairs, and later as the Vice President for Strategic Studies. In these positions, he focused on defense strategy and programs, revolutions in war, Department of Defense transformation, and maritime affairs.

Mr. Work was also an adjunct professor at George Washington University, where he taught defense analysis and roles and missions of the armed forces.

His military and civilian awards include the Legion of Merit, Meritorious Service Medal, Defense Meritorious Service Medal, and the Navy Distinguished Civilian Service Award.

He also serves on the Board of Directors at Raytheon and is the president and owner of TeamWork, LLC, which provides insight, counsel and advice on defense and national security issues.

August Cole

August Cole is an author, futurist and analyst exploring the future of conflict. He works on Creative Foresight at SparkCognition. August is a regular speaker to private sector, academic and US and allied government audiences. He is a non-resident senior fellow at the Brent Scowcroft Center on International Security at the Atlantic

Council; he directed the Council's Art of the Future Project, which explores creative and narrative works for insight into the future of conflict, from its inception in 2014 through 2017. His first book, "Ghost Fleet: A Novel of the Next World War," is a collaborative novel written with Peter W. Singer. This near-future thriller about the next world war was published in 2015. He also edited the Atlantic Council science fiction collection, War Stories From the Future, published in 2015. He is the author of numerous other short stories about the future of conflict. From 2007 to 2010, August reported on the defense industry for The Wall Street Journal and previously was an editor and reporter at MarketWatch.com. He is a regular participant in Defense Entrepreneurs Forum activities and the AI Initiative of the Future Society at the Harvard Kennedy School.

Paul Scharre

Paul Scharre is a Senior Fellow and Director of the Technology and National Security Program at the Center for a New American Security. He is the author of "Army of None: Autonomous Weapons and the Future of War."

From 2008-2013, Mr. Scharre worked in the Office of the Secretary of Defense (OSD) where he played a leading role in establishing policies on unmanned and autonomous systems and emerging weapons technologies. Mr. Scharre led the DoD working group that drafted DoD Directive 3000.09, establishing the Department's policies on autonomy in weapon systems. Mr. Scharre also led DoD efforts to establish policies on intelligence, surveillance, and reconnaissance (ISR) programs and directed energy technologies. Mr. Scharre was involved in the drafting of policy guidance in the 2012

Defense Strategic Guidance, 2010 Quadrennial Defense Review, and Secretary-level planning guidance. His most recent position was Special Assistant to the Under Secretary of Defense for Policy.

Prior to joining OSD, Mr. Scharre served as a special operations reconnaissance team leader in the Army's 3rd Ranger Battalion and completed multiple tours to Iraq and Afghanistan. He is a graduate of the Army's Airborne, Ranger, and Sniper Schools and Honor Graduate of the 75th Ranger Regiment's Ranger Indoctrination Program.

Mr. Scharre has published articles in The New York Times, Wall Street Journal, CNN, TIME, Foreign Policy, Foreign Affairs, Politico, and The National Interest, and has appeared on CNN, MSNBC, Fox News, NPR, and the BCC. He has testified before the House and Senate Armed Services Committees and has presented at the United Nations, NATO, the Pentagon, the CIA, and other national security venues. Mr. Scharre is a term member of the Council on Foreign Relations.

Prof. Bruce Porter

A two-time Chair of the University of Texas Computer Science Department, Dr. Bruce Porter serves as SparkCognition's Chief Science Officer, where he leads the company's many R&D initiatives.

Currently, as University Professor, Dr. Porter's research focuses on machine reading, a technology that holds tremendous potential for capturing knowledge for automated inference, question answering, explanation generation, and other AI capabilities.

Dr. Porter also directs UT's Knowledge Systems Research Group, an AI organization with the goal to develop methods to build knowledgeable computers. He has won the Best Paper Award at the National Conference on Artificial Intelligence, the College of Natural Sciences Teaching Excellence Award, the National Science Foundation's Presidential Young Investigator Award, and the President's Associates Teaching Excellence Award.

Wendy R. Anderson

Recognized as one of the top 25 women in the U.S. to shape defense policy, Wendy R. Anderson brings 20 years of experience in national and international security to her position as General Manager, Defense & National Security, at SparkCognition.

Anderson served in the Obama administration in key leadership positions in the Departments of Defense and Commerce: Deputy Chief of Staff to Defense Secretary Chuck Hagel, Chief of Staff to Deputy Secretary of Defense Ash Carter, and Chief of Staff to Commerce Secretary Penny Pritzker.

Previously, she held senior positions in the U.S. Senate, on both the Intelligence and Homeland Security Committees. Anderson was twice awarded the Department of Defense Medal for Distinguished Public Service, the Department's highest civilian service award. A graduate of Harvard and Columbia universities, Anderson specializes in South Asia and the Middle East. Anderson is a partner at Strong Eagle Media; an adjunct Senior Fellow in the Military, Veterans, and Society Program at the Center for a New American Security; and a member of the Council on Foreign Relations. She serves on the board of Team Rubicon Global and on the Board of Trustees at her alma mater, Hendrix College.

James Joye Townsend Jr.

Jim Townsend is an adjunct senior fellow at the Center for a New American Security (CNAS), where he co-leads the Transatlantic Security Program. He is the co-host of an interview program and writes the weekly transatlantic news feature "The Dish."

Previously, Jim worked as Deputy Assistant Secretary of Defense (DASD) for European and NATO Policy. He helped execute U.S. military engagement in almost every conflict from the Gulf War to the reintroduction of U.S. forces into Europe to deter Russia. He played critical roles in NATO enlargement, NATO reform, and building relations with the new democracies after the breakup of the Soviet Union.

Jim has also held positions as Vice President in the Atlantic Council of the United States and Director of the Council's Program on International Security, Principal Director of European and NATO Policy, Director of NATO Policy, and Director of the Defense Plans Division at the U.S. Mission to NATO in Brussels, Belgium.

Jim has been decorated by 11 European nations and multiple times by the Department of Defense, including a Presidential Rank Award (Meritorious Executive).

Index

Entries in italics indicate tables or figures respectively.

INDEX